Synthesis Lectures on Human Language Technologies

Series Editor

Graeme Hirst, Department of Computer Science, University of Toronto, Toronto, ON, Canada

The series publishes topics relating to natural language processing, computational linguistics, information retrieval, and spoken language understanding. Emphasis is on important new techniques, on new applications, and on topics that combine two or more HLT subfields.

Chen-Tse Tsai · Shyam Upadhyay · Dan Roth

Multilingual Entity Linking

Chen-Tse Tsai
Bloomberg
New York, USA

Shyam Upadhyay
Google Deepmind
New York, USA

Dan Roth
Department of Computer and Information Science
University of Pennsylvania
Pennsylvania, USA

ISSN 1947-4040 ISSN 1947-4059 (electronic)
Synthesis Lectures on Human Language Technologies
ISBN 978-3-031-74900-1 ISBN 978-3-031-74901-8 (eBook)
https://doi.org/10.1007/978-3-031-74901-8

© The Editor(s) (if applicable) and The Author(s), under exclusive license to Springer Nature Switzerland AG 2024

This work is subject to copyright. All rights are solely and exclusively licensed by the Publisher, whether the whole or part of the material is concerned, specifically the rights of translation, reprinting, reuse of illustrations, recitation, broadcasting, reproduction on microfilms or in any other physical way, and transmission or information storage and retrieval, electronic adaptation, computer software, or by similar or dissimilar methodology now known or hereafter developed.
The use of general descriptive names, registered names, trademarks, service marks, etc. in this publication does not imply, even in the absence of a specific statement, that such names are exempt from the relevant protective laws and regulations and therefore free for general use.
The publisher, the authors and the editors are safe to assume that the advice and information in this book are believed to be true and accurate at the date of publication. Neither the publisher nor the authors or the editors give a warranty, expressed or implied, with respect to the material contained herein or for any errors or omissions that may have been made. The publisher remains neutral with regard to jurisdictional claims in published maps and institutional affiliations.

This Springer imprint is published by the registered company Springer Nature Switzerland AG
The registered company address is: Gewerbestrasse 11, 6330 Cham, Switzerland

If disposing of this product, please recycle the paper.

Preface

Identifying entities and concepts, disambiguating them, and grounding them in encyclopedic resources, is a crucial step toward understanding natural language text. In this monograph, we consider the problem of grounding concepts and entities mentioned in text to one or more knowledge bases (KBs), commonly referred to as Entity Discovery and Linking (EDL). In a well-studied scenario of the EDL problem, documents are given in English and the goal is to locate concept and entity mentions, and find the corresponding entries the mentions refer to in an encyclopedic knowledge base such as Wikipedia. As most of the content on the web is not in English, there have been an increasing number of studies on identifying and grounding entities written in any language to the English Wikipedia, a task often referred to as Cross-lingual Entity Discovery and Linking.

In this book, we introduce challenges and existing solutions to both monolingual and cross-lingual EDL problems. The core challenges of both settings include how to locate important mentions that we would like to disambiguate in text, how to quickly generate plausible entities from a large KB, how to measure relevance between mentions in text and entities in a KB, and how to perform inference so that the solutions within a document are coherent.

The cross-lingual setting poses a few additional challenges beyond those addressed in the standard English EDL problem. Key among them is that no direct supervision signals are available to facilitate training machine learning models. Cross-lingual EDL introduces problems such as recognizing multilingual named entities mentioned in text, translating and transliterating non-English names into English, and computing contextual word similarity across languages. Since it is impossible to acquire manually annotated examples for all languages, building models that support cross-lingual EDL in many languages requires exploring indirect or incidental supervision signals from a range of sources, for example, multilingual KBs.

We survey relevant research papers and organize them based on their objectives and the technologies applied. We also point out resources such as popular datasets, the knowledge bases used in those datasets, and evaluation methodologies. We end with a highlight of recent trends and a discussion of future extensions of natural language grounding.

New York, USA Chen-Tse Tsai
New York, USA Shyam Upadhyay
Pennsylvania, USA Dan Roth

Acknowledgments

Chen-Tse Tsai expresses gratitude to his wife, Lue-Jane, and daughter, Celestine, for their love and unwavering support. He also thanks his colleagues at Bloomberg for their support and insightful discussions.

Shyam Upadhyay thanks his parents, wife Rohi, and son Ishaan for their love and support.

Dan Roth thanks his wife, Michal, for her love and support, and Noam, Edo, and Ella for their love and for reminding him what really matters. He also thanks his students and colleagues in the Cognitive Computation Group for their hard work over the years and for making it all possible.

January 2024

Chen-Tse Tsai
Shyam Upadhyay
Dan Roth

Contents

1 **Introduction to Entity Discovery and Linking** 1
  1.1 Introduction 1
     1.1.1 Why Entity Discovery and Linking? 2
     1.1.2 Monolingual Entity Discovery and Linking 3
     1.1.3 Entity Discovery and Linking Across Languages 4
     1.1.4 Challenges in Entity Linking 5
  1.2 Sub-Problems in Entity Discovery and Linking 6
     1.2.1 Locating Entity Mentions 7
     1.2.2 Identifying Entity Candidates 7
     1.2.3 Linking Mentions to Entities 8
  1.3 Relation to Other NLP Tasks 8
     1.3.1 Word Sense Disambiguation 8
     1.3.2 Coarse and Fine-Grained Entity Typing 9
     1.3.3 Coreference Resolution 10
  1.4 Book Outline 11
  1.5 What's Not Covered 11
  1.6 Mathematical Notation 12
  1.7 Bibliographical Notes 12
  References 13

2 **Knowledge Bases, Datasets, and Evaluation** 15
  2.1 Knowledge Bases 15
     2.1.1 Wikipedia 16
     2.1.2 Other Related Knowledge Bases 18
  2.2 Evaluation Methodology 20
  2.3 Datasets 21
     2.3.1 Evaluation Metrics 24
  2.4 Bibliographical Notes 26
  References 27

3	Overview of Entity Discovery and Linking Pipeline	31
	3.1 Mention Extraction	32
	3.2 Candidate Generation	33
	3.3 Context Sensitive Inference	33
	3.4 NIL Mention Identification	34
	3.5 NIL Mention Clustering	35
	3.6 Alternative Pipelines	35
4	**Extracting Entity Mentions**	37
	4.1 English NER	38
	4.1.1 Problem Formulation	39
	4.1.2 Models	40
	4.1.3 English NER Datasets	44
	4.1.4 English NER for EDL	45
	4.2 NER in Other Languages	47
	4.2.1 Human Annotation	47
	4.2.2 Parallel Text	48
	4.2.3 Wikipedia	50
	4.2.4 Large Multilingual Pre-trained Language Models	51
	4.2.5 Lexicons	52
	4.3 Bibliographical Notes	54
	References	55
5	**Identifying Entity Candidates**	63
	5.1 Monolingual Candidate Generation	64
	5.1.1 Dictionary Based Approaches	64
	5.1.2 Query Log Based Approaches	67
	5.1.3 Retrieval Based Approaches	67
	5.2 Cross-Lingual Candidate Generation	70
	5.2.1 Dictionary Based Approaches	71
	5.2.2 Query Log Based Approaches	72
	5.2.3 Retrieval Based Approaches	72
	5.2.4 Name Translation and Transliteration Based Approaches	74
	5.3 Discussion and Conclusions	78
	5.4 Bibliographical Notes	79
	References	80
6	**Linking Mentions to Entities**	85
	6.1 Formulating the Linking Problem	86
	6.2 Features	87
	6.2.1 Local Features	88
	6.2.2 Global Features	92

	6.3	Inference Models	93
		6.3.1 Local Inference	94
		6.3.2 Global Inference	94
	6.4	Learning Scoring Functions	96
		6.4.1 Supervised Methods	96
		6.4.2 Unsupervised Methods	99
	6.5	NIL Mention Identification	101
	6.6	Discussion and Conclusions	102
	6.7	Bibliographical Notes	103
	References		105
7	**Recent Advances and Future Directions**		111
	7.1	Recent Advances	111
		7.1.1 Dense Entity Retrieval	111
		7.1.2 Autoregressive Entity Retrieval	113
	7.2	Future Directions	115
	References		117
Appendix			119
References			135

About the Authors

Chen-Tse Tsai is a Senior Research Scientist at Bloomberg, specializing in information extraction and time series prediction. He received his Ph.D. in Computer Science from the University of Illinois Urbana-Champaign in 2017 and earned his master's degree in Computer Science from the National Taiwan University in 2011. He has authored over 20 papers presented at top-tier NLP and ML conferences, including EMNLP, NAACL, EACL, CoNLL, and AAAI. As an action editor for ACL Rolling Review and a reviewer for various NLP conferences and journals, he actively contributes to the scholarly community.

Shyam Upadhyay is a Senior Research Scientist at Google, where he has worked on products such as the Google Assistant and Bard. He received his Ph.D. from the Department of Computer and Information Science at the University of Pennsylvania in 2019. He has published over 20 papers at top-tier NLP conferences such as ACL, EMNLP, NAACL, EACL, CoNLL, and COLING. He has served as an Area Chair for various tracks at EMNLP, *SEM, AAAI, and NAACL, and is currently serving as the Associate Editor for ACM Transactions on Asian and Low-Resource Language Information Processing (TALLIP). He received his undergraduate degree in Computer Science and Engineering from the Indian Institute of Technology at Kanpur in 2013.

Dan Roth is the Eduardo D. Glandt Distinguished Professor at the Department of Computer and Information Science, University of Pennsylvania, a VP/Distinguished Scientist at Amazon AWS AI, and a Fellow of the AAAS, the ACM, AAAI, and the ACL.

In 2017, Roth was awarded the John McCarthy Award, the highest award the AI community gives to mid-career AI researchers. He was recognized "for major conceptual and theoretical advances in the modeling of natural language understanding, machine learning, and reasoning."

Roth has published broadly in machine learning, natural language processing, knowledge representation and reasoning, and learning theory, and has developed advanced machine learning-based tools for natural language applications that are being used widely. Until February 2017, he was the Editor-in-Chief of the Journal of Artificial Intelligence Research (JAIR). He has been involved in several startups; most recently, he was a Co-founder and Chief Scientist of NexLP, a startup that leverages the latest advances in Natural Language Processing (NLP), Cognitive Analytics, and Machine Learning in the legal and compliance domains. NexLP was acquired by Reveal in 2020. He received his B.A. Summa cum laude in Mathematics from the Technion, Israel, and his Ph.D. in Computer Science from Harvard University in 1995.

Introduction to Entity Discovery and Linking

1.1 Introduction

The web is a valuable source of knowledge, expressed through text written in thousands of languages. Over 40% of the content on the web is not in English, and the rate at which such content is generated has been growing dramatically.[1] A positive aspect of the web has been the growth of online, open to all, encyclopedic resources, often facilitated by collaborative efforts. The most prominent example of such an effort is Wikipedia, a high-quality on-line encyclopedia that today comprises more than 53 million articles and attracts 1.5 billion unique visitors per month. These knowledge bases are carefully curated and dynamically updated collection of documents describing entities (e.g., "Queen Victoria"), concepts (e.g., "Physics"), and events (e.g., "The Battle of Waterloo"), written and maintained by humans. Each such document has some additional information (or *metadata*) associated with it (e.g., birth location of Queen Victoria). Wikipedia, and other knowledge bases, are a rich source of human knowledge, and have become indispensable resources for knowledge acquisition and reference.

At the same time, Wikipedia is also a rich source of human language. As of January 2023, there are about 6.6 millions articles in the English Wikipedia alone, containing over 4.2 billion words.[2] Consequently, Wikipedia became an important source for learning about human language, and the natural language research community has been using it as such, inducing models that know a lot about the statistics of language use.

However, in order to advance Natural Language Understanding (NLU), our models should make use of encyclopedias or knowledge bases in a way that is closer to what humans do,

---

[1] Based on https://www.en.wikipedia.org/wiki/Languages_used_on_the_Internet.
[2] More statistics about the size of Wikipedia can be found on https://www.en.wikipedia.org/wiki/Wikipedia:Size_of_Wikipedia.

Fig. 1.1 An example of the entity discovery and linking problem. The "Alex Smith" mentioned in the two documents refer to two different entities in Wikipedia. The disambiguation page of Alex Smith from Wikipedia on the right shows that there are several athletes named Alex Smith

beyond the sheer statistics of language regularities. Human use it to acquire *knowledge* about entities, concepts and events, explore background knowledge that is relevant to the current text they read, and use it to understand concepts and relations between them. Human readers can rather easily "link" information they read and enrich it using relevant encyclopedic resources by using background knowledge and their understanding of context. Facilitating this process, of *Linking* natural language text snippets and mentions of entities, events and concepts to the relevant articles in Wikipedia, which is the topic of this book.

Consider the example in Fig. 1.1, consisting of two documents. In document (1), the spans of text "Cincinnati" and "Alex Smith" can be grounded to the relevant entries (namely, www.en.wikipedia.org/wiki/Cincinnati_Bengals and www.en.wikipedia.org/wiki/Alex_Smith_(tight_end)) in Wikipedia. In document (2), a different Alex Smith is mentioned and therefore is linked to another Wikipedia entry. This way, the entities described in the text are disambiguated by appealing to background knowledge that has been pre-compiled into a knowledge base. Without this kind of disambiguation, it could be very difficult to distinguish the two Alex Smith even for human readers.

Given a document and a knowledge base, the task of identifying entity spans in the document and grounding them to their corresponding entities in the knowledge base, is referred to as *Entity Discovery and Linking* (EDL).

1.1.1 Why Entity Discovery and Linking?

Most studies of EDL focus on locating and linking *named entities*. Therefore the name "entity" discovery and linking. Named entities are real-world objects that can be denoted with proper names, such as people, locations, organizations, etc. Named entities are usually central units of discourse, around which the content of any document is written. As a result, to understand knowledge expressed in a document, it is necessary to identify the entities mentioned in the document. Below, we discuss some applications involving language understanding that rely on EDL.

EDL is useful for automatic knowledge acquisition. Relationship between entities evolve with the occurrence of events, which are often discussed in news articles. As new information is continuously added to the Web, it is important that knowledge bases are also automatically updated. For instance, if a news article states that "the Chiefs have traded Alex Smith to another team", one should be able to extract this information from its text and update the KB accordingly. The first step to identify the entities involved in the new relationship required entity linking (i.e., grounding Chiefs to Kansas_City_Chiefs[3] and Alex Smith to Alex_Smith). That is, EDL can be a crucial step for extracting new relationships between entities and populating knowledge bases.

In addition to helping organizing new information expressed in unstructured text into structured KBs, EDL also helps in accessing information in a natural manner. For instance, EDL can be the first step in answering natural language questions pertaining to entities. To answer "Which team did Alex Smith play for before Cincinnati", it requires identifying which "Alex Smith" is being referred to in the question. Similarly, EDL also helps in identifying relevant documents in a document retrieval system. For instance, for the query "find all articles about tight end Alex Smith", a document retrieval system can fetch all documents that have linked mentions to the relevant entity Alex_Smith instead of returning all documents that contain this surface string.

From the two applications we described above, we can see that EDL enables one to "close the loop" on knowledge acquisition and retrieval. By extracting new information through grounding to KBs, one can continuously consuming the raw unstructured text from the Web, organizing it, and then retrieving it to better process future raw text. Consequently, EDL is one of the key tools to deal with the information load. We note that EDL can be extended beyond named entities, for example, grounding general concepts (e.g., physics) or event (e.g., The Battle of Waterloo). We will discuss the applications and challenges in the last chapter.

1.1.2 Monolingual Entity Discovery and Linking

Earlier work of EDL has predominantly examined the problem with respect to English documents and an English KB such as Wikipedia or Freebase. We use the term *monolingual entity discovery and linking* to refer to the setting where the language of the input document and KBs are the same.

Figure 1.1 is an example of monolingual entity linking. Using the English Wikipedia as the target KB, the names in two pieces of English text are linked to the corresponding entries

---

[3] We use the format Kansas_City_Chiefs as a shorthand of the complete Wikipedia URL, www.en.wikipedia.org/wiki/Kansas_City_Chiefs.

in the English Wikipedia. By augmenting a piece of text with such background knowledge, entity linking helps a human reader access relevant factual knowledge to complement the information expressed in the text. This not only aids understanding, but also serves as a useful component in other NLP systems.

1.1.3 Entity Discovery and Linking Across Languages

As stated earlier, a growing fraction of the web is written in languages other than English, so knowledge can be expressed in any language. *Cross-lingual entity discovery and linking* (XEDL) extends the monolingual EDL setup to the cases where the input document and the knowledge base are in different languages. The most common setting is to ground entity mentions written in *any* language to an English knowledge base (e.g., the English Wikipedia) since an English KB usually contain more information than other languages' KB.

Figure 1.2 shows an example of XEDL task where the input document is written in Tamil (a language with >70 million speakers) and the knowledge base is the English Wikipedia. The three Tamil names in the text are linked to the corresponding entries in the English Wikipedia articles. For instance, the Tamil mention லிவர்பூல் (which translates to 'Liverpool') is linked to the football club `Liverpool_F.C.`, and not the city of Liverpool or the University of Liverpool.

Besides helping to acquire and organize knowledge written in other languages automatically, cross-lingual entity linking makes this information available to a broader audience than what the language in which it was expressed can reach, helping overcome the metaphorical language-imposed barrier to knowledge. For instance, simply knowing what entities appear in a certain tweet written in Tamil can improve the understanding of an English reader.

Fig. 1.2 An example of cross-lingual entity linking. The input document is written in Tamil, and the task involves detecting named entity spans and linking them to an English Knowledge Base, in this case Wikipedia

This can help faster dissemination of critical information, that can prove crucial in urgent situations (e.g., early warning systems, disaster relief, financial trading etc.).

But why not translate the text to English? Firstly, machine translation models are not perfect. Translation errors are likely propagated to the downstream EDL task. Translating proper nouns could be especially difficult since there are more exceptions and out-of-vocabulary words. For instance, many named entities are transliterated instead of being translated according to the meanings. Secondly, building a translation system requires resources in the target language (namely large parallel corpora), which are not always available and are expensive to annotate at scale Lopez and Post (2013), especially for low-resource languages. Unless several million lines of in-domain parallel text are available, statistical machine translation approaches perform poorly, an issue that is exacerbated for the now popular neural machine translation approaches Koehn and Knowles (2017). Although recent advances (e.g. Chen et al. (2022)) improve zero-shot neural machine translation by leveraging large pre-trained multilingual language models, the performance for low-resource languages is still far from perfect. From a practical standpoint, including a translation system in the pipeline may prove to be an overkill and introduce huge latency. Another practical issue is that even though we can apply NLP tools on the translated text, for some applications, the spans of text that are grounded need to be projected back to the original document. For these problems, a translation-based approach incurs the additional cost of projecting the output back to the original language, which besides introducing latency, might be a hard task in itself.

Unlike monolingual entity linking for English, the cross-lingual entity linking is relatively less explored. The cross-lingual nature of the problem introduces new challenges to the entity linking problem. We will briefly discuss these additional challenges in Sect. 1.2, and have in-depth discussion from Chaps. 4–6.

1.1.4 Challenges in Entity Linking

At first glance, it may appear that entity linking is simply searching for a span of text against all the entities listed in a KB to find the correct grounding. However, entity linking involves addressing two key properties of languages to correctly perform grounding.

Ambiguity. Natural language often uses entity references that are *ambiguous*. That is, the same reference can refer to different entities, depending on the context in which the reference appears. For instance, in Fig. 1.1 document (2), "Alex Smith" refers to the quarterback who played for Kansas City Chiefs from 2013 to 2017 in the National Football League (NFL), in contrast to Fig. 1.1 document (1) where "Alex Smith" refers to the Cincinnati Bengals tight end. In fact, the span of text "Alex Smith" can potentially refer to over 20 entries in Wikipedia, listed in the disambiguation page www.en.wikipedia.org/wiki/Alex_Smith_ (disambiguation) (shown on the right in Fig. 1.1). Human readers might easily identify that the two "Alex Smith"s in this example refer to different entities since "Cincinnati" and

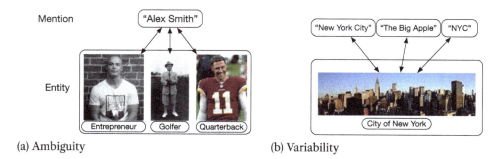

Fig. 1.3 Ambiguity and **Variability**: The two key challenges in Entity Linking

"Chiefs" are likely to be two different teams. However, it is crucial for an EDL system to perform a similar inference which utilizes the context around the mention to ground these mentions properly.

Variability. A related challenge is that of *variability*. References to the same entity can be made in more than one way. For instance, the entity New_York_City can also be referred to as "the Big Apple", "NYC", or "The Five Boroughs". Often, the number of ways to refer to the same entity is correlated with its popularity on the web. According to Wikipedia,[4] there are about 12 common nicknames for New_York_City besides several acronyms. To address this challenge, a common solution is to obtain a pre-compiled list of names that can refer to a given entity. Such lists usually can be extracted from the knowledge base.

Figure 1.3 summarizes the two main challenges, ambiguity and variability. Dealing with these challenges in EDL requires appealing to both contextual information present in the text and background knowledge expressed in a KB.

1.2 Sub-Problems in Entity Discovery and Linking

Entity Discovery and Linking is usually decoupled into the following sub-problems: locating entity mentions, identifying entity candidates, and linking mentions to entities. Each of these sub-problems could be a standalone research topic. We briefly introduce these sub-problems in this section, and will have in-depth discuss in Chaps. 4–6.

---

[4] https://www.en.wikipedia.org/wiki/Nicknames_of_New_York_City.

1.2.1 Locating Entity Mentions

The first and foremost challenge is about identifying words which are part of names.[5] This research topic is usually referred as the mention extraction problem. For instance, in a sentence

> Sunday's Super Bowl is between the Denver Broncos and Seattle Seahawks.

The string "Sunday's Super Bowl" could be wrongly recognized as a named entity instead of "Super Bowl" since "Sunday" is also capitalized. On the other hand, a system may only identify "Denver" instead of the full name "Denver Broncos" since Denver is a more well-known location name. Texts from social medias or discussion forums could be more challenging than the ones from news articles since they often contain more nicknames and abbreviations, and do not have proper capitalization.

In the cross-lingual scenario, the problem becomes identifying entity mentions in a piece of text written in a non-English language. For the languages which do not use space to separate words, such as Chinese and Japanese, the mention extraction problem could be much more difficult. Moreover, for low-resource languages which do not have enough resources for building good machine learning models, this first sub-problem becomes very challenging.

Mention extraction is often viewed as a separate problem. Many entity linking works assume that entity mentions are given as part of the input, and therefore only focus on linking the given entity mentions to entries in a KB.

1.2.2 Identifying Entity Candidates

After named entity mentions are extracted, the next problem is to identify plausible entities in the given KB that the mention might be referring to. Identifying plausible candidates is challenging due to *ambiguity* and *variability* of natural language.

Since a knowledge base usually contains more than millions of records, the key question here is how to quickly reduce the number of possible entities to a manageable size, so that a more sophisticated and resource-hungry algorithm can be applied to disambiguate these candidates. Take the mention Alex Smith in Fig. 1.1 as an example, instead of comparing "Alex Smith" with millions of entries in the English Wikipedia, it makes more sense to only consider the entries that could be referred to as Alex Smith, such as the entities listed in the disambiguation page of Alex Smith.

---

[5] We assume the goal is to disambiguate named entities in this section. The same idea can be applied on disambiguating concepts or events.

In the cross-lingual setup, there is an additional challenge of how to compare mentions in one language with entities in another language. For instance, in Fig. 1.2, candidates from the English Wikipedia have to be proposed for the Tamil surface form லிவர்பூல் that translates to Liverpool.

1.2.3 Linking Mentions to Entities

Given the extracted mentions and the corresponding entity candidates, the next problem is to choose the correct entities from the candidate sets.

The key challenge in this problem is to measure relevancy between mentions in the text and entities in a knowledge base. Context words or other mentions in the document are usually very important for differentiating similar names. For instance, to ground the two Alex Smith mentions in Fig. 1.1, the team names in the two documents ("Cincinnati" and "The Chiefs") are the most important clues that disambiguate the two mentions. Consequently, in order to achieve correct disambiguation, an EDL system needs to understand the context, and also leverages information described in the knowledge base.

1.3 Relation to Other NLP Tasks

As disambiguation is critical for understanding natural languages, besides entity linking, several NLP tasks also aim at disambiguating or understanding words and entities in text. We briefly discuss three related tasks in this section: word sense disambiguation, coarse and fine-grained named entity typing, and coreference resolution.

1.3.1 Word Sense Disambiguation

Natural language is ambiguous: words can have more than one distinct meaning. For instance, consider the following sentences:

> I can't hear the bass guitar.
> I like to go bass fishing.

The word *bass* clearly has different meanings in the two sentences. The correct sense of an ambiguous word depends on the context in which it occurs. The problem of word sense disambiguation (WSD) is the task of assigning the correct meaning to a polysemous word within a given context. WSD is a classical task in the field of NLP. It was convinced as an essential task for machine translation in the late 1940s Weaver (1952). Ide and Veronis (1998) present more in-depth history of WSD.

A sense inventory provides the senses a word can be associated with. Earlier works on the WSD problem focus on disambiguating words using thesauri or machine-readable dictionaries as the sense inventories. Later, WordNet (Miller et al. 1990) becomes the most widely used sense inventory for WSD. More recently, Mihalcea (2007) considers Wikipedia and uses the hyperlinked text in Wikipedia articles to generate annotated training data automatically. Other encyclopedia, such as BabelNet (Moro et al. 2014), also has been used for multilingual WSD.

A key difference between entity linking and WSD is that entity linking usually focuses on named entity mentions where as WSD focuses on common nouns. The mention is complete in WSD, but in entity linking, the mention is potentially incomplete since an entity can be expressed in various ways, such as abbreviations or aliases. For instance, "Alex Smith" can be referred only by "Alex". Therefore, the mention extraction step is more challenging for entity linking. Moreover, in WSD the candidate senses for a word are provided in an inventory. Due to potentially incomplete mentions and huge ambiguity of name entities, entity linking systems usually have an additional step—retrieving entity candidates from millions of records based on mention surface strings.

In principle, both EL and WSD deal with the ambiguity problem in natural language. However, named entities are often ambiguous, whereas most words assume only a single sense (i.e., monosemous) Hachey et al. (2013). In this respect, entity linking could be a harder disambiguation task than WSD.

1.3.2 Coarse and Fine-Grained Entity Typing

Named entity recognition (NER) is the task of identifying and typing named entity mentions. The commonly used types are *Person*, *Location*, and *Organization*. For example, given the following sentence:

> U.N. official Rolf Ekeus heads for Baghdad

the goal is to identify that "U.N." is an organization, "Rolf Ekeus" is a person, and "Baghdad" is a location. Besides locating entity mentions, the task also requires to assign an entity type to each mention. Some datasets also include a couple of more types such as *Facility*, *Nationality*, *Product*, …etc. These types are usually viewed as coarse-grained entity types. Since these coarse types may not disambiguate entities well, researchers have proposed to use fine-grained types instead. For instance, Ling and Weld (2012) created a dataset with 112 unique tags based on FreeBase types, and the dataset created by Choi et al. (2018) contains 10,201 ultra fine-grained types based on common noun phrases.

Conceptually, Entity Discovery and Linking can be viewed as an extremely fine-grained NER problem. Namely, instead of using few common entity types, entity linking uses millions of types (each entity entry in the KB is a type). In contrast, NER performs disambiguation at a very coarse level. It can tell the difference between "Washington" as a person and as a location, but it could not distinguish Washington (State) from Washington (Park) since both of them will be tagged as a location. Even with the fine-grained typing taxonomy, the two "Alex Smith" in Fig. 1.1 may be both labeled as *person.athlete*, therefore could not be differentiated properly.

Solutions for entity typing usually do not rely on an external knowledge base, but only focus on performing disambiguation using the input text. However, a good entity linking system can potentially help to improve entity typing since it is relatively easy to map entries in a KB to the typing taxonomy used in an entity typing problem. For instance, Zhou et al. (2018) leverages entity linking for zero-shot fine-grained entity typeing.

The ability to identifying named entity mentions of NER is aligned with the need of EDL. Therefore, NER is usually used as the mention detector in an EDL system. We will introduce and discuss more details about NER in Chap. 4.

1.3.3 Coreference Resolution

Coreference resolution is the task of finding all spans of text that refer to the same entity in a given document. For instance, in the document in Fig. 1.4 identifying that $[\text{queen}]_{12}$ and $[\text{she}]_{13}$ refer to the same entity, namely `Queen_Elizabeth_II`. Formally, coreference resolution is a clustering problem where mentions refer to the same entity are put into one cluster.

The key difference between entity linking and coreference resolution is that coreference only performs disambiguation within a given document. It provides information on whether two mentions in the document refer to the same entity, but it does not try to ground the corresponding entity in a knowledge base. For instance, in the example document in Fig. 1.4, coreference resolution places the mention "Laura Ingraham" in a cluster {5} all by itself, while EDL grounds it to `Laura_Ingraham` in a KB (such as Wikipedia), allowing the reader to access more information about the entity, if needed.

Nevertheless, coreference resolution and entity linking are mutually beneficial. In principle, if one has a perfect cross-document coreference resolution system one can solve EDL.

On Thursday, a day after [President Trump's]$_1$ first state visit to [Britain]$_2$ ended, [he]$_3$ raved in an interview with the [Fox News]$_4$ host [Laura Ingraham]$_5$. For [his]$_6$ state visit, [Trump]$_7$ brought along not only the first lady, [Melania Trump]$_8$, but also [his]$_9$ four adult children. [Trump]$_{10}$ said that [he]$_{11}$ had developed a deep bond with the [queen]$_{12}$ and that [she]$_{13}$ had been very taken with [him]$_{14}$.

Fig. 1.4 An example text with entity mentions marked. Coreference resolution clusters the entity mentions as follows: {1, 3, 6, 7, 9, 10, 11, 14}, {12, 13}. That is, mentions {12, 13} all refer to the same entity

For such reduction, the input to the coreference system would be the input text to EDL and the text for all entities in the KB. Similarly, a perfect EDL system can solve coreference by grounding every mention and then placing all mentions that ground to the same entity in a coreference cluster. Often EDL systems attempt to exploit the document-internal coreference relationships to aggregate contextual information to improve EDL. Similarly, EDL has been used for improving coreference resolution Ratinov and Roth (2012). Additionally, there have been approaches that attempt to perform EDL and coreference resolution jointly.

1.4 Book Outline

Chapter 3 discusses the EDL pipeline and the resources used in building EDL systems. In particular, we review different knowledge bases, (such as Wikipedia) and evaluation datasets and metrics. The chapter goes into details of the structure of Wikipedia, a commonly used KB, and how its inter-lingual structure can be leveraged for building a cross-lingual EDL system. Chapters 4, 5, and 6 discuss each of the main components of the EDL pipeline in detail.

In Chap. 4, we will introduce the named entity recognition problem, a popular mention detection approaches for both monolingual and cross-lingual EDL.

Chapter 5 discusses different approaches for generating entity candidates for the extracted mentions. We categorize candidate generation methods into three categories: dictionary based, query-log based, and retrieval based approaches for both monolingual and cross-lingual entity linking. We will also discuss approaches based on name translation and transliteration for dealing with additional challenges in the cross-lingual setup.

Chapter 6 discusses the final inference step: selecting the best entity from the candidate set. We will first introduce common features that measure similarity or compatibility between mentions in the input document and entities in the KB, local and global inference approaches, and the models that learn functions to score candidate entities.

We will highlight recent advances in EDL research and provide future research directions in Chap. 7.

Since EDL is an active research topic, the main content of this manuscript will cover works as of mid 2022. Most related papers published in 2022 will only be discussed in the bibliographical note sections at the end of each chapter.

1.5 What's Not Covered

While we aim to provide a comprehensive overview of EDL and cross-lingual EDL, describing in detail some of the building blocks used in an EDL system is beyond the scope of this book. For instance, Chap. 4 deals with mention detection problem. We only describe most common ways to perform named entity recognition in a document since this is not a book about NER. In Chap. 6, we will discuss approaches to represent the context of a mention. A key ingredient in this chapter involves word embeddings—both monolingual and multilingual. We will briefly introduce popular word representations in Appendix A.1 and context representations in A.2. For a detailed description of word monolingual and multilingual embeddings, please refer to Søgaard et al. (2019). Machine learning models like SVMs, RankSVM, and CRF, and inference techniques such as Belief Propagation and Integer Linear Programming will also be introduced in the Appendix. However, we do not cover details of every models that have been applied to EDL.

1.6 Mathematical Notation

We will use italicized lowercase letters $x, y, z, \ldots$ to denote scalars, bold lowercase letters $\mathbf{v}, \mathbf{w}$ to denote vectors, bold uppercase letters $\mathbf{V}, \mathbf{W}$ to denote matrices, and [] to index entries of vectors or matrices. That is, $\mathbf{v}[i]$ is the ith element of the vector $\mathbf{v}$ and $\mathbf{W}[i][j]$ is the entry in the ith row and jth column of the matrix $\mathbf{W}$. The operator arg max $\mathbf{v}$ identifies the index of the element with the maximum value in a vector $\mathbf{v}$. The dot product of a vector $\mathbf{v}$ with another vector $\mathbf{w}$ is denoted using $\mathbf{v}^T \mathbf{w}$, and $\mathbf{W}\mathbf{v}$ denotes the product of a matrix $\mathbf{W}$ and vector $\mathbf{v}$. A sequence of vectors $\mathbf{v}_s, \mathbf{v}_{s+1}, \mathbf{v}_{s+2}, \mathbf{v}_{s+n}$ is denoted by $\mathbf{v}_{s:s+n}$. Sets are denoted using italicized uppercase letters A, B and set intersection and union are denoted by $A \cap B$ and $A \cup B$ respectively. The probability of an event s is denoted using $\Pr(s)$, and of event s conditioned on event t is denoted by $\Pr(s \mid t)$.

We use m to denote a mention, $\mathcal{D}$ to denote a document, and $\mathcal{K}$ to denote a knowledge base. One or more mentions m_1, m_2 can belong to a document $m_1, m_2 \in \mathcal{D}$. A knowledge base $\mathcal{K}$ is a set of entities $\{e_1, e_2, e_3, \ldots, e_i, \ldots\}$ where e_i denotes an entity. For a mention m, the gold disambiguating entity is denoted using e^*, and the predicted entity is denoted using $\hat{e}$.

1.7 Bibliographical Notes

Earlier work on entity linking has predominantly examined the problem with respect to English documents and an English KB like Wikipedia or Freebase (Bunescu and Pasca 2006; Cheng and Roth 2013; Hoffart et al. 2011; Mihalcea and Csomai 2007; Ratinov et al. 2011; Shen et al. 2015, inter alia).

Entity Discovery and Linking has been shown useful in several NLP applications, such as question answering Khalid et al. (2008), Sun et al. (2015), Yih et al. (2015), coreference resolution Hajishirzi et al. (2013), Ponzetto and Strube (2006), Rahman and Ng (2011), knowledge base population Ji and Grishman (2011), and multiparty dialogues understanding Chen and Choi (2016).

References

Bunescu, R., Paşca, M.: Using encyclopedic knowledge for named entity disambiguation. In: 11th Conference of the European Chapter of the Association for Computational Linguistics, pp. 9–16. Association for Computational Linguistics, Trento, Italy (2006). https://aclanthology.org/E06-1002

Chen, Y.-H., Choi, J.D.: Character identification on multiparty conversation: Identifying mentions of characters in TV shows. In: Proceedings of the 17th Annual Meeting of the Special Interest Group on Discourse and Dialogue, pp. 90–100. Association for Computational Linguistic, Los Angeles (2016). https://doi.org/10.18653/v1/W16-3612. https://aclanthology.org/W16-3612

Chen, G., Ma, S., Chen, Y., Zhang, D., Pan, J., Wang, W., Wei, F.: Towards making the most of cross-lingual transfer for zero-shot neural machine translation. In: Proceedings of the 60th Annual Meeting of the Association for Computational Linguistics (Volume 1: Long Papers), pp. 142–157. Association for Computational Linguistic, Dublin, Ireland (2022). https://doi.org/10.18653/v1/2022.acl-long.12. https://aclanthology.org/2022.acl-long.12

Cheng, X., Roth, D.: Relational inference for wikification. In: Proceedings of the 2013 Conference on Empirical Methods in Natural Language Processing, pp. 1787–1796. Association for Computational Linguistics, Seattle, Washington, USA (2013). https://aclanthology.org/D13-1184

Choi, E., Levy, O., Choi, Y., Zettlemoyer, L.: Ultra-fine entity typing. In: Proceedings of the 56th Annual Meeting of the Association for Computational Linguistics (Volume 1: Long Papers), pp. 87–96. Association for Computational Lingui, Melbourne, Australia (2018https://doi.org/10.18653/v1/P18-1009. https://aclanthology.org/P18-1009

Hachey, B., Radford, W., Nothman, J., Honnibal, M., Curran, J.R.: Evaluating entity linking with wikipedia. Artificial intelligence **194**, 130–150 (2013)

Hajishirzi, H., Zilles, L., Weld, D.S., Zettlemoyer, L.: Joint coreference resolution and named-entity linking with multi-pass sieves. In *Proceedings of the 2013 Conference on Empirical Methods in Natural Language Processing*, pages 289–299, Seattle, Washington, USA, Oct. 2013. Association for Computational Linguistics. https://aclanthology.org/D13-1029

Hoffart, J., Yosef, M.A., Bordino, I., Fürstenau, H., Pinkal, M., Spaniol, M., Taneva, B., Thater, S., Weikum, G.: Robust disambiguation of named entities in text. In: Proceedings of the 2011 Conference on Empirical Methods in Natural Language Processing, pp. 782–792. Association for Computational Linguistics, Edinburgh, Scotland, UK (2011). https://aclanthology.org/D11-1072

Ide, N., Véronis, J.: Introduction to the special issue on word sense disambiguation: The state of the art. Computational Linguistics **24**(1), 1–40 (1998). https://aclanthology.org/J98-1001

Ji, H., Grishman, R.: Knowledge base population: Successful approaches and challenges. In: Proceedings of the 49th Annual Meeting of the Association for Computational Linguistics: Human Language Technologies, pp. 1148–1158. Association for Computational Linguistics, Portland, Oregon, USA (2011). https://aclanthology.org/P11-1115

Khalid, M.A., Jijkoun, V., De Rijke, M.: The impact of named entity normalization on information retrieval for question answering. In: Proceedings of the European Conference on Information Retrieval (ECIR), pp. 705–710. Springer (2008)

Koehn, P., Knowles, R.: Six challenges for neural machine translation. In: Proceedings of the First Workshop on Neural Machine Translation, pp. 28–39. Association for Computational Linguistics, Vancouver (2017). https://doi.org/10.18653/v1/W17-3204. https://aclanthology.org/W17-3204

Ling, X., Weld, D.S.: Fine-grained entity recognition. In: Proceedings of the National Conference on Artificial Intelligence (AAAI) (2012). http://aiweb.cs.washington.edu/ai/pubs/ling-aaai12.pdf

Lopez, A., Post, M.: Beyond bitext: Five open problems in machine translation. In: Proceedings of the EMNLP Workshop on Twenty Years of Bitext (2013)

Mihalcea, R., Csomai, A.: Wikify!: Linking documents to encyclopedic knowledge. In: Proceedings of the ACM Conference on Information and Knowledge Management (CIKM) (2007)

Mihalcea, R.: Using Wikipedia for automatic word sense disambiguation. In: Human Language Technologies 2007: The Conference of the North American Chapter of the Association for Computational Linguistics; Proceedings of the Main Conference, pp. 196–203. Association for Computational Linguistics, Rochester, New York (2007). https://aclanthology.org/N07-1025

Miller, G., Beckwith, R., Fellbaum, C., Gross, D., Miller, K.: Wordnet: An on-line lexical database. International Journal of Lexicography (1990)

Moro, A., Raganato, A., Navigli, R.: Entity linking meets word sense disambiguation: a unified approach. Transactions of the Association for Computational Linguistics **2**, 231–244 (2014). https://doi.org/10.1162/tacl_a_00179

Ponzetto, S.P., Strube, M.: Exploiting semantic role labeling, WordNet and Wikipedia for coreference resolution. In: Proceedings of the Human Language Technology Conference of the NAACL, Main Conference, pp. 192–199. Association for Computational Linguistics, New York City, USA (2006). https://aclanthology.org/N06-1025

Rahman, A., Ng, V.: Coreference resolution with world knowledge. In: Proceedings of the 49th Annual Meeting of the Association for Computational Linguistics: Human Language Technologies, pp. 814–824. Association for Computational Linguistics, Portland, Oregon, USA (2011). https://aclanthology.org/P11-1082

Ratinov, L., Roth, D., Downey, D., Anderson, M.: Local and global algorithms for disambiguation to Wikipedia. In: Proceedings of the 49th Annual Meeting of the Association for Computational Linguistics: Human Language Technologies, pp. 1375–1384. Association for Computational Linguistics, Portland, Oregon, USA (2011). https://aclanthology.org/P11-1138

Ratinov, L., Roth, D.: Learning-based multi-sieve co-reference resolution with knowledge. In: Proceedings of the 2012 Joint Conference on Empirical Methods in Natural Language Processing and Computational Natural Language Learning, pp. 1234–1244. Association for Computational Linguistics, Jeju Island, Korea (2012). https://aclanthology.org/D12-1113

Shen, W., Wang, J., Han, J.: Entity linking with a knowledge base: Issues, techniques, and solutions. IEEE Transactions on Knowledge and Data Engineering **27**(2), 443–460 (2015)

Søgaard, A., Vulić, I., Ruder, S., Faruqui, M.: Cross-lingual word embeddings. Synthesis Lectures on Human Language Technologies **12**(2), 1–132 (2019)

Sun, H., Ma, H., Yih, W.-T., Tsai, C.-T., Liu, J., Chang, M.-W.: Open domain question answering via semantic enrichment. In: Proceedings of the International World Wide Web Conference (WWW), pp. 1045–1055. International World Wide Web Conferences Steering Committee (2015)

Weaver, W.: Translation. In: Proceedings of the Conference on Mechanical Translation, Massachusetts Institute of Technology, pp. 17-20 (1952). https://aclanthology.org/1952.earlymt-1.1

Yih, W.-T., Chang, M.-W., He, X., Gao, J.: Semantic parsing via staged query graph generation: Question answering with knowledge base. In: Proceedings of the 53rd Annual Meeting of the Association for Computational Linguistics and the 7th International Joint Conference on Natural Language Processing (Volume 1: Long Papers), pp. 1321–1331. Association for Computational Linguistic, Beijing, China (2015). https://doi.org/10.3115/v1/P15-1128. https://aclanthology.org/P15-1128

Zhou, B., Khashabi, D., Tsai, C.-T., Roth, D.: Zero-shot open entity typing as type-compatible grounding. In: Proceedings of the 2018 Conference on Empirical Methods in Natural Language Processing, pp. 2065–2076. Association for Computational Linguistic, Brussels, Belgium (2018). https://doi.org/10.18653/v1/D18-1231. https://aclanthology.org/D18-1231

2 Knowledge Bases, Datasets, and Evaluation

In this chapter, we delve into the data and evaluation methodologies associated with the entity discovery and linking problem (EDL). A crucial input for this task is the target knowledge base. Some knowledge bases feature rich textual descriptions of entities, complemented by a plethora of attributes and relationships among these entities. In contrast, others might simply provide a brief description for each listed entity. There are knowledge bases that encompass entities articulated in multiple languages, while some are tailored to a specific language or domain. The content within a knowledge base dictates the challenges faced when anchoring entities to varying bases.

In addition to discussing mainstream knowledge bases, this chapter will also touch upon training and test data. The quality and volume of annotated data often steer the selection of machine learning models and their subsequent performance. It's noteworthy that datasets might operate under varied assumptions and present distinct definitions of the problem at hand. Grasping the intricacies of these resources is pivotal, ensuring their optimal utilization and enabling a nuanced comparison of research findings.

Lastly, machine learning models are inherently metric-driven. A model is typically designed to optimize a predetermined loss function. The comparison between different models is usually based on evaluation metrics derived from the test segments of datasets. To fully appreciate this research domain, it's indispensable to understand the metrics that gauge its advancements before delving into solutions and intricate technicalities.

2.1 Knowledge Bases

A knowledge base is essential for entity linking, defining both the scope of entities and the information available for disambiguation. At its core, a knowledge base must have clear, unambiguous entries that correspond to real-world entities or specific concepts in a domain. Furthermore, each entry in a knowledge base typically includes various properties or attributes, such as its definition, description, category, and relationships with other entries.

In this section, we'll delve into Wikipedia, the premier knowledge base favored by the research community for entity linking. Its popularity stems from its expansive domain coverage, detailed information within and across entities, multilingual capabilities, interlanguage links, and rigorous quality control and verification processes. Such features have enabled researchers to experiment with and validate a plethora of concepts. In Sect. 2.1.2, we'll explore other prominent knowledge bases utilized for entity linking, many of which are intertwined with Wikipedia, including DBpedia, Freebase, and Wikidata.

2.1.1 Wikipedia

Wikipedia[1] is a free online multilingual encyclopedia with the aim to allow anyone to edit it. It is the most popular Internet encyclopedia in the world and is a very quickly growing resource. As in August 2023, there are 6.7 million articles in the English Wikipedia.[2] When the target knowledge base is Wikipedia, the entity linking task is also referred as *Wikification* (Milne and Witten 2008; Ratinov et al. 2011).

An entry in Wikipedia is an article which can be referred by an unique *title* (e.g., Barack_Obama[3]) or identifier (e.g., Page ID[4]: 534366). The article describes important information of the entity. Moreover, several phrases in this article could be hyperlinked to the corresponding entries in Wikipedia. For instance, in the first paragraph of Barack Obama's Wikipedia page, the phrase "African American" is linked to Wikipedia page African_Americans, and "U.S. Senator" is linked to the page United_States_Senate. These hyperlinked text are usually referred as *anchor text*. They are created by the users with an intention to help readers to comprehend the article easier. Figure 2.1 shows an example Wikipedia page and the corresponding information page.

Every article in Wikipedia is required to have at least one *category*. For instance, some categories of Barack Obama's page are "Presidents of the United States", "Illinois Democrats", and "University of Chicago Law School faculty". Categories allow articles to be placed into one or more topics. These topics can be further categorized by one or more parent categories.

Another useful structure is the *disambiguation pages*, which are created for ambiguous names. For example, CIA_(disambiguation) is a disambiguation page which contains about 30 entities which can be referred by "CIA". Some examples are Cairo_International_Airport, California_Institute_of_the_Arts, and Certified_Internal_Auditor.

---

[1] https://en.wikipedia.org/.
[2] https://en.wikipedia.org/wiki/Wikipedia:Size_of_Wikipedia.
[3] When expressing Wikipedia titles, we use underscores instead of white spaces.
[4] Found on https://en.wikipedia.org/w/index.php?title=Barack_Obama&action=info.

2.1 Knowledge Bases

Fig. 2.1 An example of Wikipedia page and its metadata. From the left figure, we can see that there are several blue links in the article which introduces Barack Obama. These anchor texts point to the corresponding entities or concepts in Wikipedia. In the beginning of this page, there is a sentence stating that `Barack` and `Obama` are both redirected to this page. In the right figure, several information of the page is listed, such as Page ID, number of redirects, and the Wikidata ID

There is a special kind of pages called *redirect*. A redirect page exists for each alternative name of entities in Wikipedia. For example, "America", "US", "U.S.", and "USA" are some of the redirect pages for the Wikipedia page `United_States`. Accessing the redirect pages will bring users to the page of the target entry (United_States). Redirects are very useful in normalizing entity names in text, and are often used in the candidate generation step of an EDL pipeline.

Wikipedias exist in over 250 languages. Concepts described in different Wikipedias often overlap, indirectly introducing information redundancy across languages, as some concepts have articles in more than one language. Such articles in Wikipedia are linked through *inter-language links*. Formally, an inter-language link $\mathcal{A} \leftrightarrow \mathcal{B}$ indicates that article $\mathcal{A}$ in the language l_1's Wikipedia and article $\mathcal{B}$ in language l_2's Wikipedia describe the same entity or concept.

For instance, Fig. 2.2 shows that the article अल्बर्ट_आइंस्टीन in the Hindi Wikipedia corresponds to the article `Albert_Einstein` in the English Wikipedia. These inter-language links identify the same concept in different Wikipedias, thereby unifying the concepts across languages. Table 2.1 shows the statistics of the number of articles in Wikipedias for 9 languages, along with the size of the intersection of the respective Wikipedia with the articles in English Wikipedia. We can see that the coverage of inter-language links is far from perfect since some entities may only exist in one language. Even an entity exists in two languages of Wikipedia, the inter-language link between them may still be missing since it is created manually.

From Table 2.1, we also see that the size of Wikipedia of different languages vary a lot. The German Wikipedia has more than 2 millions articles, whereas there are less than 100 thousands articles in the Tagalog Wikipedia. The size could impact the quality of models in several steps of the pipeline. For instance, it is common to use Wikipedia articles to

Fig. 2.2 Inter-language links in English Wikipedia and Hindi Wikipedia link the corresponding articles that describe the entity `Albert_Einstein` in both Wikipedias

Table 2.1 Statistics showing number of articles in Wikipedias for 9 languages, and the size of the intersection (as identified by inter-language links) with the English Wikipedia. The last column shows the relative size of the intersection (in %). For instance, 52% of German Wikipedia articles have an English Wikipedia counterpart

Language	# Articles	Intersection	%
German	2.06M	1.06M	52
French	1.87M	1.21M	65
Italian	1.36M	927k	68
Spanish	1.29M	850k	66
Chinese	942k	519k	55
Arabic	521k	330k	63
Turkish	292k	203k	69
Tamil	104k	61k	59
Tagalog	69k	53k	77

train word and entity representations (embeddings). If a language's Wikipedia is small, its monolingual or cross-lingual embeddings are expected to have lower quality in general.

2.1.2 Other Related Knowledge Bases

In this section, we will briefly introduce other common knowledge bases which have been used in the Entity Linking research. Many of them are actually derived or extended from Wikipedia. In-depth comparison between some of them can be found in Färber et al. (2015).

2.1 Knowledge Bases

DBpedia (Auer et al. 2007). DBpedia[5] is a multilingual knowledge base constructed by extracting structured information from various Wikimedia projects, such as Wikipedia's infoboxes, categories, geo-coordinates, and the links to external web pages. The 2022 December English snapshot of DBpedia contains 850 million facts (triples), including 1.7 million persons, 748 thousands places, 346 thousand organizations, 1.9 million species, and 10 thousands diseases.[6] Since DBpedia is organized in database structure, it is easier to obtain relations between entities using DBpedia than parsing the Wikipedia dump. Several entity linking research using DBPedia are along the line of DBpedia Spotlight[7] (Mendes et al. 2011; Chabchoub et al. 2018; Sakor et al. 2019).

YAGO (Suchanek et al. 2007; Rebele et al. 2016). YAGO[8] was one of the first academic projects to build a knowledge base automatically. YAGO was mainly built from Wikipedia, WordNet, and GeoNames. As of 2019, it contains more than 10 million entities and 120 million facts about these entities. YAGO focuses on extracting high quality information from 10 different languages of Wikipedia. The extractions are regularly assessed by the YAGO team manually. YAGO also attaches a temporal dimension and a spatial dimension to many of its entities and facts, which allow users to query the data over space and time.

YAGO4 (Pellissier Tanon et al. 2020) was proposed in 2020. Instead of combing Wikiepdia and WordNet, it collects items in Wikidata. In addition, it uses rigorous type hierarchy with semantic constraints. The complex taxonomy of Wikidata is replaced by the simpler and clean taxonomy of schema.org (Guha et al. 2016). This refined ontology is logically consistent, facilitating semantic reasoning with OWL description logics.

Freebase (Bollacker et al. 2008). Freebase is a large knowledge base which was collaboratively created mainly by its community members. It is an online collection of structured data harvested from many sources such as Wikipedia, NNDB, and MusicBrainz. As of 2014, Freebase had approximately 44 million instances and 2.4 billion facts.

The typing information in Freebase is considered much cleaner and more structured than the categories in Wikipedia. The typing taxonomy is organized as two layers of hierarchy. The first layer contains coarse types such as `organization`, `location`, and `sports`. Under the first layer `location` type, some examples of the second layer types are `country`, `citytown`, or `us_state`. It is straightforward to map Freebase types to the commonly used entity types for named entity recognition. This automatic type mapping allows creation of NER training data and features for entity linking. Since Freebase entries are linked to Wikipedia entries, one can use the Freebase types even when the target knowledge base is Wikipedia.

---

[5] http://wiki.dbpedia.org/.
[6] https://www.dbpedia.org/blog/dbpedia-snapshot-2022-12-release/.
[7] https://www.dbpedia-spotlight.org/.
[8] https://yago-knowledge.org/.

A snapshot of Freebase is used as the target knowledge base[9] for the entity discovery and linking task in Text Analysis Conference (TAC) from 2015 (Ji et al. 2015) to 2018. In 2015, Freebase was officially shut down, although the database dump is still available online.[10] It is replaced by the Wikidata project.[11]

WikiData (Vrandečić 201). Wikidata is a collaborative knowledge base which provides structured information. Unlike text-based Wikipedia, each item in Wikidata is described by property-value pairs. Each property can include a list of references to the sources that support the claim. The aim of Wikidata is to provide data which can be used by other knowledge bases such as Wikipeida. Wikidata acts as the central storage for the structured data of Wikimedia projects including Wikipedia, Wiktionary, Wikisource, and others. As of 2023, there are 106 million items in Wikidata.

WikiData is multilingual in the sense that entries also contains names and aliases in several languages, and they are linked to the corresponding Wikipedia pages in different languages. An increasing amount of Entity Linking research use WikiData as the target knowledge base for both monolingual and cross-lingual setups. Example entity linking systems which link against Wikidata include Delpeuch (2019), Mulang et al. (2019), and Botha et al. (2020).

Biomedical Knowledge Bases. Besides general purpose knowledge bases, there are several KBs or ontologies in the biomedical domain. To name a few, Protein Ontology,[12] Sequence Ontology,[13] Chemical Entities of Biological Interest Ontology,[14] Cell Ontology,[15] and Gene Database.[16] Several of these ontologies are integrated into Unified Medical Language System (UMLS) (Bodenreider 2004), the largest single ontology of biomedical concepts. These ontologies have been used to generate entity linking datasets and systems for the biomedical domain.

2.2 Evaluation Methodology

Datasets and automatic evaluation methodologies drive the direction of machine learning research. Annotations in datasets are usually the main signals for supervised learning models. Datasets also allow various ideas to be compared fairly and efficiently. In this section,

---

[9] The dataset ID on https://www.ldc.upenn.edu/ is LDC2015E42.
[10] https://developers.google.com/freebase/.
[11] https://www.wikidata.org.
[12] https://proconsortium.org/.
[13] http://www.sequenceontology.org/.
[14] https://www.ebi.ac.uk/chebi/.
[15] http://www.obofoundry.org/ontology/cl.html.
[16] https://www.ncbi.nlm.nih.gov/gene.

we will first discuss the key datasets used in the EDL literature, and then introduce various evaluation measures. The less popular or newer datasets will be listed in the bibliographical notes and briefly discussed at the end of this chapter.

2.3 Datasets

Annotations of linking mentions to entries of a knowledge base can be acquired manually or automatically. In the manual workflow, annotators are hired to provide labels for a collection of documents. Usually, there are clear annotation guidelines which defines the task and resolves edge cases. Many researchers also leverage the web to obtain entity linking labels automatically. For instance, the hyperlinked text in Wikipedia articles provide lots of free annotations. Authors of general web pages or blogs could also sometimes link text to Wikipedia entries. If these automatically obtained labels are only used for training a model, one potential problem is that the domain of these documents could be different from the evaluation domain. On the other hand, the quality of labels usually needs to be checked in order to be used as an evaluation data. Another common issue is that not all mentions are linked. For instance, only the first mention of each entity is linked in Wikipedia articles. Therefore, these automatically generated datasets are usually not suitable for training or evaluating the mention extraction component.

Another issue of older datasets is that Wikipedia titles are used as the labels instead of Wikipedia Page IDs. This could be a problem if an EDL system uses a different version of Wikipedia. Since Wikipedia titles may be changed over time, the gold titles in these datasets may not exist in a newer version of Wikipedia.

In the following, we list popular datasets that have been used for training or evaluating EDL models. Table 2.2 further summarizes and compares these datasets. Newer datasets and datasets for specific domains are listed in the bibliographical notes at the end of this chapter.

AQUAINT (Milne and Witten 2008) contains 727 mentions which are linked to the English Wikipedia. The documents are from the AQUAINT text corpus: a collection of newswire stories from the Xinhua News Service, the New York Times, and the Associated Press. This dataset is annotated to mimic the hyperlink style of Wikipedia. Namely, only the first mentions of important entities are annotated.

MSNBC (Cucerzan 2007) contains 756 mentions which are linked to the English Wikipedia. The source documents are from MSNBC news. The mentions are extracted by running an NER and coreference resolution system. In this dataset, all the detected mentions are annotated, including some NIL annotations.

ACE (Ratinov et al. 2011) contains 257 mentions. This is a subset of the ACE coreference data set (Doddington et al. 2004). The authors leverage the gold typing and coreference information to annotate the corresponding Wikipedia titles using Amazon Mechanical Turk.

Wikipedia (Ratinov et al. 2011) contains 928 mentions from randomly selected 40 paragraphs in Wikipedia. Since most anchor texts can be easily resolved by the popularity features, the authors try to make this data set more challenging by removing most of the easy mentions. Note that several works (Bunescu and Paşca 2006; Milne and Witten 2008; Cucerzan 2007) have also created their own evaluation data using Wikipedia. However, this data set is made publicly available and is used in several studies.

AIDA-CoNLL (Hoffart et al. 2011) contains roughly 35k mentions (including training, development, and test sets) from Reuters news articles. This dataset was originally created for CoNLL 2003 English NER shared task (Tjong Kim Sang and De Meulder 2003). The paper authors hand-annotated all mentions with corresponding entries in YAGO2, Freebase, and Wikipedia. Since the named entity mentions are annotated exhaustively, this dataset is suitable for evaluating both NER and entity linking.

TAC (Text Analysis Conference—Knowledge Base Population) hosts EL and EDL shared tasks from 2009 to 2018 (Getman et al. 2018). The National Institute of Standards and Technology (NIST) releases new corpus for evaluation and/or training every year. The numbers of newly annotated mentions every year are listed in Table 2.2. Datasets provided in the previous years are usually included in training and developing models. The source documents are from three broad domains: news, discussion forums, and weblogs. Besides EDL annotations, the mentions are also labeled with coarse named entity types. TAC KB is built from the October 2008 dump of English Wikipedia, which is used as the target knowledge base until 2014. Starting from 2015, mentions are linked to BaseKB, a snapshot of Freebase.

Starting from 2014, English named entity mentions are annotated exhaustively, therefore these datasets are also suitable for evaluating NER performance. Starting from 2011, Chinese mention queries are included for evaluating cross-lingual entity linking performance, whereas Spanish mention queries are added since 2012. These cross-lingual entity linking datasets drive early development of cross-lingual entity linking research.

LORELEI (Tracey et al. 2019; Tracey and Strassel 2020) The DARPA (Defense Advanced Research Projects Agency) LORELEI Program[17] aims to improve the performance of NLP technologies in the context of providing situation information for a natural disaster or other emergent incident, with a particular focus on low resource languages. Through the course of the program, datasets for 20 representative languages and 9 incident languages were released. These datasets contain resources such as parallel text, comparable text, semantic annotations, and entity annotations. The full test sets of incident languages and 25k words for each representative language have entity linking annotations. An English knowledge

---

[17] https://www.nist.gov/itl/iad/mig/lorehlt-evaluations.

2.3 Datasets

Table 2.2 Common datasets for EDL and cross-lingual EDL. The number of mentions considers both training and test sets if applicable. The "Exhaustive?" column indicates if all mentions are annotated. A dataset with exhaustively annotated mentions is suitable for evaluating mention extraction. The 26 languages contained in the LORELEI dataset which have EL annotations are Akan, Amharic, Arabic, Bengali, English, Farsi, Hindi, Hungarian, Ilocano, Indonesian, Kinyarwanda, Mandarin, Odia, Oromo, Russian, Sinhala, Spanish, Swahili, Tagalog, Tamil, Thai, Tigrinya, Ukrainian, Vietnamese, Wolof, and Zulu. [†]25k words per language are annotated for entity linking, and additional 75k words are labeled for NER. Both TAC and LORELEI datasets can be found on LDC website: https://catalog.ldc.upenn.edu/byproject

Data Set	# Mentions	Exhaustive?	Source	KB	Languages
AQUAINT	727	No	News	Wikipedia	English
MSNBC	756	No	News	Wikipedia	English
ACE	257	Yes	News, Broadcast	Wikipedia	English
Wikipedia	928	No	Wikipedia	Wikipedia	English
AIDA-CoNLL	34,956	Yes	News	Wikipedia, YAGO, Freebase	English
TAC 2009	3,904	No	News	TAC KB	English
TAC 2010	3,750	No	News, Web	TAC KB	English
TAC 2011	2,250	No	News, Web	TAC KB	English
	4,338				Chinese
TAC 2012	2,229	No	News, Web, Broadcast	TAC KB	English
	2,180				Chinese
	3,916				Spanish
TAC 2013	2,190	No	News, Web, Discussion forum	TAC KB	English
	2,155				Chinese
	2,117				Spanish
TAC 2014	11,020	Yes	News, Web, Discussion forum	TAC KB	English
	3,253	No			Chinese
	2,598	No			Spanish

(continued)

Table 2.2 (continued)

Data Set	# Mentions	Exhaustive?	Source	KB	Languages
TAC 2015	29,190	Yes	News, Discussion forum	BaseKB	English
	24,182				Chinese
	9,999				Spanish
TAC 2016	8,845	Yes	News, Discussion forum	BaseKB	English
	9,231				Chinese
	6,964				Spanish
TAC 2017	6,915	Yes	News, Discussion forum	BaseKB	English
	10,246				Chinese
	7,212				Spanish
LORELEI	25k words†	Yes	News, Blogs, Social media, Discussion forum	GeoNames, CIA World Leaders List, CIA World Factbook	26 languages

base is constructed from GeoNames,[18] CIA World Leaders List,[19] Appendix B of the CIA World Factbook,[20] and manually augmented incident-relevant entities.

2.3.1 Evaluation Metrics

In this section, we introduce common measurements for assessing the quality of model outputs. Different metrics are used for different problem setups, for example, if the mentions are given or if the NIL mention clustering is applied.

Bag-of-Title F_1. In the earlier studies (Milne and Witten 2008; Ratinov et al. 2011) in which Wikipedia titles are used to represent entities, Bag-of-Title F_1 (BOT) is used to evaluate entity linking systems when the mentions are given at test time. Given a document, let the set of predicted entities be P, and the set of gold entities be G. We have

---

[18] http://www.geonames.org/.
[19] https://www.cia.gov/library/publications/world-leaders-1/.
[20] https://www.cia.gov/library/publications/resources/the-world-factbook/appendix/appendix-b.html.

2.3 Datasets

$$\text{precision} = \frac{|P \cap G|}{|P|}, \text{ and recall} = \frac{|P \cap G|}{|G|},$$

where $\cap$ indicates set intersection. The F_1 score is the harmonic mean of precision and recall:

$$F_1 = 2 \times \frac{\text{precision} \times \text{recall}}{\text{precision} + \text{recall}}$$

A drawback of using BOT metric is that it does not evaluate the connection between mentions and the linked entities. That is, it only considers what entities appear in a document, but does not know which mentions corresponds to which entities. As a result, a system can achieve a high BOT score without any correct prediction. For example, suppose there are two mentions in a document: "Alex Smith" and "Chiefs", and their gold Wikipedia pages are Alex_Smith and Kansas_City_Chiefs respectively. If a system links Alex Smith to Kansas_City_Chiefs and Chiefs to Alex_Smith, it still gets a perfect BOT F_1 score since these two entities both exist in the gold entity set.

This metric is good for capturing topical information of a document, but does not directly evaluate the ability of disambiguating each mention.

F_1 score. A more common evaluation metric for entity linking systems is the mention level F_1 score. Each predicted and gold mention is represented as a tuple of three fields: (start offset, end offset, knowledge base ID or NIL), where start and end character offsets specify the mention location in the document. A predicted mention is considered matched with a gold mention if all three fields in the tuple are identical to the ones of the gold mention. Using this definition, micro or macro F_1 score can be calculated across all documents.

For the systems do not perform mention extraction and only focus on disambiguating gold mentions, the knowledge base ID is the only evaluated field since start and end offsets are given. In this case, the F_1 score will be identical to precision and recall, which is also known as the *precision@1* or *accuracy* measure. Moreover, instead of only taking the top candidate into account, the *precision@k* measure evaluates if the answer is within the top k candidates.

Coreference metrics. When the NIL mention clustering step is applied, we can view the output of an entity linking system as a result of cross-document coreference resolution. Mentions which are linked to the same knowledge base entry are clustered together. Namely, the knowledge base ID of a mention is used as the cluster ID. And the NIL mentions are also clustered based on system's NIL mention clustering algorithm.

In this case, several evaluation metrics for coreference resolution can be applied, including MUC (Vilain et al. 1995), B-cubed (Bagga and Baldwin 1998), and CEAF (Luo 2005). These metrics are all included in the official evaluation script[21] of TAC-KBP entity linking tasks. The metric Mention CEAF is considered as one of the major end-to-end evaluation measures in the shared tasks.

---

[21] https://github.com/wikilinks/neleval.

Given gold clusters $\{G_1, \ldots, G_m\}$ and predicted clusters $\{P_1, \ldots, P_n\}$, CEAF metric first computes an alignment between the two clustering. Given the constraint that one cluster can only be aligned at most once, this maximal bipartite matching problem can be solved in polynomial time using Kuhn-Munkres Algorithm (Kuhn 1955). A common similarity metric between two clusters is $\phi(G_i, P_i) = |G_i \cap P_i|$. The CEAF precision and recall are defined as follows:

$$\text{precision} = \frac{\Phi}{\sum_i \phi(P_i, P_i)}, \text{ and recall} = \frac{\Phi}{\sum_i \phi(G_i, G_i)},$$

where Φ represents the sum of similarities between the aligned cluster pairs. We note that there are different variants of CEAF measures (Cai and Strube 2010) which better handle the unaligned clusters.

2.4 Bibliographical Notes

Detailed description of several years of TAC-KBP datasets can be found in McNamee and Dang (2009), Ji et al. (2010, 2011, 2014, 2015), and (2016). Getman et al. (2018) summarize nine years of linguistic resources for TAC-KBP. Besides the datasets we introduced in Sect. 2.2, following datasets have also been used in some research papers.

English EDL Datasets IITB (Kulkarni et al. 2009) contains about 17k mentions from 107 web pages which are in the domains of sports, entertainment, science and technology, and health. The goal is to have high recall annotations, therefore the human annotators were told to be as exhaustive as possible. KORE50 (Hoffart et al. 2012) contains 144 mentions from 50 short English sentences. The idea is to build a data set with high ambiguity. Logeswaran et al. (2019) build a dataset[22] for zero-shot entity linking. The source documents and entity annotations are taken from FANDOM,[23] a community-written encyclopedia specializing in subjects such as video games, animations, or film series. More recently, Joko et al. (2021) created ConEL,[24] an entity linking dataset for conversational systems. Disambiguating concepts and named entities in dialogues is important for understanding the intent of user utterances.

Several datasets have been created for the biomedical or biological areas. CRAFT (Bada et al. 2012) (The Colorado Richly Annotated Full-Text Corpus[25]) contains 97 articles from the PubMed Central Open Access Subset. The biomedical concepts in these documents are linked to several ontologies in the field such as Cell Ontology, Gene Ontology, Molecular Process Ontology, Protein Ontology, and NCBI Taxonomy. Basaldella et al. (2020) create

---

[22] https://github.com/lajanugen/zeshel.
[23] https://www.fandom.com/.
[24] https://github.com/informagi/conversational-entity-linking.
[25] https://github.com/UCDenver-ccp/CRAFT.

COMETA, a corpus consisting of 20,000 English biomedical entity mentions from Reddit posts. Mentions are linked to SNOMED CT,[26] a medical knowledge graph, by domain experts. The concepts in COMETA include symptoms, diseases, anatomical expressions, chemicals, genes, devices and procedures across a range of conditions.

Cross-lingual EDL Datasets For the cross-lingual entity linking setting, McNamee et al. (2011) create a corpus[27] of 21 non-English languages for cross-lingual entity linking. The person names in the non-English documents are linked to the TAC KB (in English). The English documents in the parallel corpora are annotated by an English NER model and linked using Amazon's Mechanical Turk. Person name mentions are then projected to the non-English documents via word alignments automatically.

Tsai and Roth (2016b) create a dataset consists of Wikipedia articles from 12 non-English languages of Wikipedia. The hyperlinked texts are treated as the mentions and the corresponding English Wikipedia titles are obtained via the cross-language links in Wikipedia. Only the mentions which have English Wikipedia titles are kept. Since most mentions in Wikipedia articles are easy, in the sense that some popularity feature could disambiguate them correctly, they make the number of easy mentions about twice as many as the hard mentions.

Botha et al. (2020) propose Mewsli-9, a corpus contains 289,087 entity mentions from 58,717 news articles from WikiNews with links to WikiData. The corpus includes documents in nine diverse languages. This dataset features many entities that lack English Wikipedia pages. About 11% of the distinct entities do not have English Wikipedia pages.

References

Auer, S., Bizer, C., Kobilarov, G., Lehmann, J., Cyganiak, R., Ives, Z.: DBpedia: a nucleus for a web of open data. In: The Semantic Web, pp. 722–735 (2007)

Bada, M., Eckert, M., Evans, D., Garcia, K., Shipley, K., Sitnikov, D., Baumgartner, W.A., Cohen, K.B., Verspoor, K., Blake, J.A., Hunter, L.E.: Concept annotation in the CRAFT corpus. BMC Bioinform. (2012)

Bagga, A., Baldwin, B.: Entity-based cross-document coreferencing using the vector space model. In: COLING 1998 Volume 1: The 17th International Conference on Computational Linguistics (1998). https://aclanthology.org/C98-1012

Basaldella, M., Liu, F., Shareghi, E., Collier, N.: COMETA: a corpus for medical entity linking in the social media. In: Proceedings of the 2020 Conference on Empirical Methods in Natural Language Processing (EMNLP), pp. 3122–3137. Association for Computational Linguistics (2020). https://doi.org/10.18653/v1/2020.emnlp-main.253. URL https://aclanthology.org/2020.emnlp-main.253

Bodenreider, O.: The unified medical language system (UMLS): integrating biomedical terminology. Nucl. Acids Res. (2004)

---

[26] https://www.snomed.org/snomed-ct/why-snomed-ct.

[27] https://hltcoe.jhu.edu/research/datasets-and-resources/.

Bollacker, K., Evans, C., Paritosh, P., Sturge, T., Taylor, J.: Freebase: a collaboratively created graph database for structuring human knowledge. In: Proceedings of the 2008 ACM SIGMOD International Conference on Management of Data. ACM (2008)

Botha, J.A., Shan, Z., Gillick, D.: Entity Linking in 100 Languages. In: Proceedings of the 2020 Conference on Empirical Methods in Natural Language Processing (EMNLP), pp. 7833–7845. Association for Computational Linguistics (2020). https://doi.org/10.18653/v1/2020.emnlp-main.630. URL https://aclanthology.org/2020.emnlp-main.630

Bunescu, R., Paşca, M.: Using encyclopedic knowledge for named entity disambiguation. In: 11th Conference of the European Chapter of the Association for Computational Linguistics, pp. 9–16. Association for Computational Linguistics, Trento, Italy (2006). https://aclanthology.org/E06-1002

Cai, J., Strube, M.: Evaluation metrics for end-to-end coreference resolution systems. In: Proceedings of the SIGDIAL 2010 Conference, pp. 28–36. Association for Computational Linguistics, Tokyo, Japan (2010). https://aclanthology.org/W10-4305

Chabchoub, M., Gagnon, M., Zouaq, A.: FICLONE: improving DBpedia spotlight using named entity recognition and collective disambiguation. Open J. Semant. Web (OJSW) **5**(1), 12–28 (2018)

Cucerzan, S.: Large-scale named entity disambiguation based on Wikipedia data. In: Proceedings of the 2007 Joint Conference on Empirical Methods in Natural Language Processing and Computational Natural Language Learning (EMNLP-CoNLL), pp. 708–716. Association for Computational Linguistics, Prague, Czech Republic (2007). https://aclanthology.org/D07-1074

Delpeuch, A.: Opentapioca: lightweight entity linking for wikidata. arXiv:1904.09131, 2019

Doddington, G., Mitchell, A., Przybocki, M., Ramshaw, L., Strassel, S., Weischedel, R.: The automatic content extraction (ACE) program–tasks, data, and evaluation. In: Proceedings of the Fourth International Conference on Language Resources and Evaluation (LREC'04). European Language Resources Association (ELRA), Lisbon, Portugal (2004). http://www.lrec-conf.org/proceedings/lrec2004/pdf/5.pdf

Färber, M., Ell, B., Menne, C., Rettinger, A.: A comparative survey of DBpedia, Freebase, OpenCyc, Wikidata, and YAGO. Semant. Web J. **1**(1), 1–5 (2015)

Getman, J., Ellis, J., Strassel, S., Song, Z., Tracey, J.: Laying the groundwork for knowledge base population: nine years of linguistic resources for TAC KBP. In: Proceedings of the Eleventh International Conference on Language Resources and Evaluation (LREC 2018). European Language Resources Association (ELRA), Miyazaki, Japan (2018). https://aclanthology.org/L18-1245

Guha, R.V., Brickley, D., Macbeth, S.: Schema. org: evolution of structured data on the web. Commun. ACM **59**(2), 44–51 (2016)

Hoffart, J., Yosef, M.A., Bordino, I., Fürstenau, H., Pinkal, M., Spaniol, M., Taneva, B., Thater, S., Weikum, G.: Robust disambiguation of named entities in text. In: Proceedings of the 2011 Conference on Empirical Methods in Natural Language Processing, pp. 782–792. Association for Computational Linguistics, Edinburgh, Scotland, UK (2011). https://aclanthology.org/D11-1072

Hoffart, J., Seufert, S., Nguyen, D.B., Theobald, M., Weikum, G.: KORE: keyphrase overlap relatedness for entity disambiguation. In: Proceedings of the ACM Conference on Information and Knowledge Management (CIKM), pp. 545–554. ACM (2012)

Ji, H., Grishman, R., Dang, H.T., Griffitt, K., Ellis, J.: Overview of the TAC 2010 knowledge base population track. In: Text Analysis Conference (TAC) (2010)

Ji, H., Grishman, R., Dang, H.T.: Overview of the TAC2011 knowledge base population track (2011)

Ji, H., Nothman, J., Hachey, B.: Overview of TAC-KBP2014 entity discovery and linking tasks. In: Text Analysis Conference (TAC) (2014)

Ji, H., Nothman, J., Hachey, B., Florian, R.: Overview of TAC-KBP2015 tri-lingual entity discovery and linking. In: Text Analysis Conference (TAC) (2015)

Ji, H., Nothman, J., Dang, H.T., Hub, S.I.: Overview of TAC-KBP2016 tri-lingual EDL and its impact on end-to-end cold-start KBP (2016)

Joko, H., Hasibi, F., Balog, K., de Vries, A.P.: Conversational entity linking: problem definition and datasets. In: Proceedings of the 44th International ACM SIGIR Conference on Research and Development in Information Retrieval, pp. 2390–2397 (2021)

Kuhn, H.W.: The Hungarian method for the assignment problem. Nav. Res. Logist. Q. **2**(1–2), 83–97 (1955)

Kulkarni, S., Singh, A., Ramakrishnan, G., Chakrabarti, S.: Collective annotation of Wikipedia entities in web text. In: Proceedings of the 15th ACM SIGKDD Conference on Knowledge Discovery and Data Mining (KDD), pp. 457–466. ACM (2009)

Logeswaran, L., Chang, M.-W., Lee, K., Toutanova, K., Devlin, J., Lee, H.: Zero-shot entity linking by reading entity descriptions. In: Proceedings of the 57th Annual Meeting of the Association for Computational Linguistics, pp. 3449–3460. Association for Computational Linguistics, Florence, Italy (2019). https://doi.org/10.18653/v1/P19-1335. https://aclanthology.org/P19-1335

Luo, X.: On coreference resolution performance metrics. In: Proceedings of Human Language Technology Conference and Conference on Empirical Methods in Natural Language Processing, pp. 25–32. Association for Computational Linguistics, Vancouver, British Columbia, Canada (2005). https://aclanthology.org/H05-1004

McNamee, P., Dang, H.T.: Overview of the TAC 2009 knowledge base population track. In: Text Analysis Conference (TAC), vol. 17, pp. 111–113 (2009)

McNamee, P., Mayfield, J., Lawrie, D., Oard, D., Doermann, D.: Cross-language entity linking. In: Proceedings of 5th International Joint Conference on Natural Language Processing, pp. 255–263. Asian Federation of Natural Language Processing, Chiang Mai, Thailand (2011). https://aclanthology.org/I11-1029

Mendes, P.N., Jakob, M., García-Silva, A., Bizer, C.: DBpedia spotlight: shedding light on the web of documents. In: Proceedings of the 7th International Conference on Semantic Systems, pp. 1–8. ACM (2011)

Milne, D., Witten, I.H.: Learning to link with Wikipedia. In: Proceedings of the ACM Conference on Information and Knowledge Management (CIKM) (2008)

Mulang, I.O., Singh, K., Vyas, A., Shekarpour, S., Sakor, A., Vidal, M.E., Auer, S., Lehmann, J.: Context-aware entity linking with attentive neural networks on wikidata knowledge graph (2019)

Pellissier Tanon, T., Weikum, G., Suchanek, F.: YAGO 4: a reason-able knowledge base. In: The Semantic Web: 17th International Conference, pp. 583–596. Springer (2020)

Ratinov, L., Roth, D., Downey, D., Anderson, M.: Local and global algorithms for disambiguation to Wikipedia. In: Proceedings of the 49th Annual Meeting of the Association for Computational Linguistics: Human Language Technologies, pp. 1375–1384. Association for Computational Linguistics, Portland, Oregon, USA (2011). https://aclanthology.org/P11-1138

Rebele, T., Suchanek, F., Hoffart, J., Biega, J., Kuzey, E., Weikum, G.: YAGO: a multilingual knowledge base from Wikipedia, Wordnet, and Geonames. In: International Semantic Web Conference, pp. 177–185. Springer (2016)

Sakor, A., Singh, K., Vidal, M.: FALCON: an entity and relation linking framework over DBpedia. In: Proceedings of the ISWC 2019 Satellite Tracks, pp. 265–268 (2019)

Suchanek, F.M., Kasneci, G., Weikum, G.: YAGO: a core of semantic knowledge. In: Proceedings of the International World Wide Web Conference (WWW) (2007)

Tjong Kim Sang, E.F., De Meulder, F.: Introduction to the CoNLL-2003 shared task: Language-independent named entity recognition. In: Proceedings of the Seventh Conference on Natural Language Learning at HLT-NAACL 2003, pp. 142–147 (2003). https://aclanthology.org/W03-0419

Tracey, J., Strassel, S.: Basic language resources for 31 languages (plus English): the LORELEI representative and incident language packs. In: Proceedings of the 1st Joint Workshop on Spoken Language Technologies for Under-resourced languages (SLTU) and Collaboration and Computing for Under-Resourced Languages (CCURL), pp. 277–284. European Language Resources association, Marseille, France (2020). ISBN 979-10-95546-35-1. https://aclanthology.org/2020.sltu-1.39

Tracey, J., Strassel, S., Bies, A., Song, Z., Arrigo, M., Griffitt, K., Delgado, D., Graff, D., Kulick, S., Mott, J., Kuster, N.: Corpus building for low resource languages in the DARPA LORELEI program. In: Proceedings of the 2nd Workshop on Technologies for MT of Low Resource Languages, pp. 48–55. European Association for Machine Translation, Dublin, Ireland (2019). https://aclanthology.org/W19-6808

Tsai, C.-T., Roth, D.: Cross-lingual wikification using multilingual embeddings. In: Proceedings of the 2016 Conference of the North American Chapter of the Association for Computational Linguistics: Human Language Technologies, pp. 589–598. Association for Computational Linguistics, San Diego, California (2016b). https://doi.org/10.18653/v1/N16-1072. https://aclanthology.org/N16-1072

Vilain, M., Burger, J., Aberdeen, J., Connolly, D., Hirschman, L.: A model-theoretic coreference scoring scheme. In: Sixth Message Understanding Conference (MUC-6): Proceedings of a Conference Held in Columbia, Maryland, November 6–8, 1995, (1995). https://aclanthology.org/M95-1005

Vrandečić, D.: Wikidata: a new platform for collaborative data collection. In: Proceedings of the International World Wide Web Conference (WWW), pp. 1063–1064. ACM (2012)

Overview of Entity Discovery and Linking Pipeline 3

Having gained an understanding of datasets, knowledge bases, and evaluation metrics, we are now poised to delve into solutions for Entity Discovery and Linking (EDL). In Chap. 2, we addressed several challenges involved in EDL, such as locating entity mentions, identifying potential entity candidates, and linking these mentions to actual entities. Consequently, EDL is often broken down into smaller, more manageable tasks and approached in a pipeline fashion. Each stage of the pipeline focuses on solving a specific sub-problem within EDL, thereby simplifying the tasks that follow.

A common framework for solving entity discovery and linking problem consists of steps such as mention extraction, candidate entity generation, context sensitive inference, NIL mention identification, and NIL mention clustering. Figure 3.1 illustrates the pipeline with a toy example shown below each step. This section provides an overview of the pipeline. Each component of the pipeline will be discussed briefly in the following sub-sections. Since most research on EDL focus on the first three steps of the pipeline, we will then expand them into the following three chapters. In the end of this section, we will also introduce alternative approaches which do not follow the pipeline framework strictly.

It's worth noting that in the academic literature, research papers frequently concentrate on just one stage of this pipeline. These studies often assume that the outputs from preceding stages are available as inputs. For example, it is a common practice to assume that the entity mentions within a document have already been identified, leaving the task reduced to disambiguating these mentions by linking them to the most appropriate entities in a knowledge base.

In this chapter, we will introduce a common framework for tackling the EDL challenge. We'll explore the motivations and hurdles associated with each component of the pipeline, considering both monolingual and cross-lingual contexts. Subsequent chapters will provide

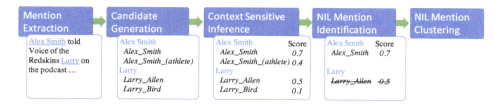

Fig. 3.1 A common entity discovery and linking pipeline consists of five components: mention detection, candidate generation, context sensitive inference, NIL mention identification, and NIL mention clustering. The last two steps are often considered optional

a detailed discussion of each sub-problem, expanding on the concepts introduced here. Although the pipeline framework we present is a common approach to solving EDL, it's not the only one. Alternative methods that address multiple sub-problems in a joint manner also exist and will be briefly touched upon at the end of this chapter and in chapters to follow.

3.1 Mention Extraction

The first step is to identify mentions (phrases in text) which we would like to disambiguate. For instance, "Alex Smith" and "Larry" are the mentions identified in the first step of Fig. 3.1. Different applications may have different definitions of mention. Mentions could be proper nouns, common nouns, pronouns, or any key-phrases that are explained in a knowledge base. Moreover, mentions could be nested or have overlap. In the example of Fig. 3.1, "Voice of the Redskins" and "Redskins" could both be mentions. Grounding both of them may be useful for some downstream applications.

Most research only focus on named entities because names usually carry the key information of the text and they could be highly ambiguous. In addition, since Named Entity Recognition is a well-studied problem, it is often used as the mention extractor in an EDL pipeline. Mention extraction is usually formulated as a sequence tagging problem and solved by training a machine learning model on lots of annotated examples. However, high-quality supervision is only available for very few languages in the world. When it comes to multilingual applications, the availability of good quality training data becomes a crucial challenge for this step.

Since mention extraction is the first step of an EDL pipeline, the quality of the extracted mention may greatly affect the performance of an entity linking system. For instance, if a model only extracts partial name "Alex" instead of "Alex Smith", it would be much more challenging for the following steps to disambiguate this name correctly.

3.2 Candidate Generation

The second step is to generate a set of entity candidates for each extracted mention. The goal of this step is to quickly reduce the number of possible entities to a manageable size, so that a more sophisticated and resource-hungry algorithm can be applied to disambiguate these candidates. Take the mention "Alex Smith" in Fig. 3.1 as an example, instead of comparing Alex Smith with millions of entries in the entire English Wikipedia, it makes more sense to only consider the entries that could be referred as "Alex Smith", such as the entities listed in the disambiguation page of Alex Smith. For the example in Fig. 3.1, two Wikipedia entities (titled `Alex_Smith` and `Alex_Smith_(athlete)`) are retrieved as the candidates for the mention "Alex Smith".

The key question of this step is how many candidates to generate so that the correct answer is included in the candidate set. There is a trade-off here: too many candidates may make this disambiguation problem more challenging, so the model in the following step suffers; on the other hand, the correct entity may not be included in the candidate set if we only retrieve a small number of candidates.

Earlier approaches to generate candidates are usually not context sensitive. Namely, other words in the sentence or document are not used in this process. In this case, lists of aliases or nicknames are the key source of information. The fundamental problem in this step is computing name similarity between the extracted mentions and names in the knowledge base. The intuition is that the target entity is unlikely to have a name (or alias) that is very different from what is mentioned in the text. In order to achieve fast retrieval, hashing techniques are applied to index all the possible names and alias in the knowledge base. Several context-sensitive candidate generation methods have been proposed recently. The key idea is that the mention representation is not static, instead, it depends on the sentence in which the mention appears.

For cross-lingual entity linking, candidate generation becomes much more challenging than the monolingual scenario since the extracted mention and the names in the knowledge base are in different languages. To be able to reuse the index built from an English knowledge base, a common strategy is to translate or transliterate the foreign mention into English, and then apply candidate generation algorithm as in the monolingual setting. Furthermore, cross-lingual word representations, which have gained significant popularity in recent years, have also been used to improve the candidate generation step for cross-lingual entity linking problem.

3.3 Context Sensitive Inference

Given the extracted mentions and the corresponding entity candidates, the next step is to assign a score to each candidate which indicates how relevant the entity is to the mention. The candidate with the highest relevance score will be picked as the output of this step. This step is usually viewed as a ranking problem: we would like to choose the best option from the candidate set.

The key challenge in this step is how to measure relevancy between mentions in the text and entities in a knowledge base. That is, how to better represent the mentions in the input document and the entities in the knowledge base, so that the score generated based on these representations can achieve disambiguation. Context words and other entities mentioned in the document are usually very important clues for differentiating similar names. For the example mention "Alex Smith" in Fig. 3.1, the context word "Redskins" might be more relevant to `Alex_Smith` than to `Alex_Smith_(athlete)`, therefore the former will be scored higher if this context information is used properly. We can see that in order to achieve this, a model needs to use information described in the knowledge base well. For example, the English Wikipedia page `Alex_Smith` specifies the fact that Alex was in Washington Redskins from 2018 to 2020.

Another challenge in this sub-problem is that mentions in a document could be interdependent. If both "Alex Smith" and "Redskins" are the mentions we would like to disambiguate, choosing the right entity for "Alex Smith" depends on if "Redskins" is linked correctly, and vice versa. Ideally, we would like to make a joint decision for all mentions in a document. However, the inference problem will quickly become infeasible as the number of mentions grows. This challenge is usually be dealt with by simplifying the dependency structure among mentions or by using approximate inference techniques.

For cross-lingual entity linking, the challenge becomes generating semantic representation across languages since the entities in knowledge bases are described in English whereas the input document is written in a foreign language. Several cross-lingual word and entity embedding approaches have been applied to tackle this challenge.

3.4 NIL Mention Identification

The coverage of the target knowledge base is usually not perfect. Entities or concepts written in text may not exist in the given knowledge base. Usually, only popular or prominent entities will be added to a knowledge base. This situation could be profounder for the domain of social media in which any topic could be discussed.

The goal of this step is to determine if the top-ranked candidate is actually the correct answer to a mention. If the target concept or entity does not exist in the knowledge base, the answer proposed by the context sensitive inference step should be rejected. These unlinkable mentions are often referred as *NIL mentions* in the literature. In the running example of Fig. 3.1, the broadcaster of Washington Redskins, Larry Michael, does not have a Wikipedia page, so the mention should not be linked to the top-ranked candidate, `Larry_Allen`.

A simple solution to this problem is to employ a threshold on the scores produced by the previous step. More specifically, if the relevance score of the top entity candidate is lower than the threshold, "NIL" will be returned instead. Another common approach is to learn a supervised binary classifier to decide if the top candidate should be rejected.

Most entity linking studies simply do not handle this issue and assume that the target entities are always in the knowledge base. In this case, only the mentions have no entity candidate are NIL. We will discuss some references which deal with this problem together with the Context Sensitive Inference step in Chap. 6 (Sect. 6.4.2.2).

3.5 NIL Mention Clustering

In the application of knowledge base population in which the goal is to enrich a knowledge base using the information extracted from text, there is an additional NIL mention clustering step. The idea is that these unlinkable mentions are potentially new entities which can be added into a knowledge base. In order to gather more accurate and comprehensive information (e.g., relations with other entities) of these NIL mentions, it is useful to cluster the NIL mentions based on the actual entities they refer to. This problem is essentially the cross-document coreference resolution problem, which is out of focus of this book.

3.6 Alternative Pipelines

Besides the EDL pipeline discussed above, one alternative framework which has been investigated by several researchers is to perform mention extraction and disambiguation jointly. More specifically, the most well-studied setup is doing Named Entity Recognition and Entity Linking together since these two tasks may reinforce each other. On one hand, the entity candidates may be constrained by the predicted entity type from an NER model. For instance, if we know that "Alex Smith" must be a person in the example of Fig. 3.1, we can ignore candidate entities which do not refer to a person therefore simplify the problem. One the other hand, entities in the knowledge base may provide hints for mention boundaries and entity types. For instance, a model would know that "Alex Smith" could be a more complete mention than simply "Alex" if multiple entities are called Alex Smith in the knowledge base.

A common idea of these joint models is to over-generate mention candidates and let the model perform disambiguation or reject the mention candidates simultaneously. We will discuss more technical details of these models in Chap. 4.

Extracting Entity Mentions 4

The first step of the entity linking pipeline is to locate phrases in text which we would like to disambiguate. These phrases are usually called *mentions* of entities or concepts. Mention extraction could be very challenging depending on the availability of resources. It is a critical component to the EDL pipeline since errors in this step are likely to be propagated into the following steps.

Figure 4.1 shows an example of input and output of the mention extraction component. In this example, we follow the most common definition of mentions for entity linking, so only the two people names are labeled in the output. However, the definition of mentions has been task dependent. While named entities or proper nouns are the most widely targeted mention type, there has been work also try to link nominal nouns or any phrase that could be linked to the knowledge base since a KB may contain concepts other than named entities. Take the input sentence in Fig. 4.1 as an example, the string "Voice of the Redskins" could be grounded to Wikipedia page `List_of_Washington_Football_Team_broadcasters` or even the current broadcaster, `Bram_Weinstein`, in 2020. Similarly, the string "podcast" could also be linked to the Wikipedia page `Podcast`. In fact, earlier works on entity linking do not restrict mentions to be named entities (Mihalcea and Csomai 2007; Milne and Witten 2008). Rather, they view mention extraction as a key phrase extraction problem where mentions are anything that can be linked to Wikipedia.

Despite the intuitive notion of 'mention', the definition of mentions has been task dependent; consequently, this has influenced the work on EDL. In this chapter, we will focus on the named entity recognition (NER) problem since it is the most widely used mention extractor for entity linking. We will discuss issues of moving beyond named entities in Chap. 7. This chapter is started with the well-studied English NER problem, including problem formulation, non-neural and neural network models, and annotated corpora. We will then introduce

Fig. 4.1 An example of the input and output of mention detection component

how English NER is extended to other languages, with more focus on the low-resource scenario in which it is hard to obtain human supervision for the target languages. In the bibliographical notes at the end of this chapter, we will have some discussion regarding the earlier work that do not use NER for extracting mentions.

We note that many studies on entity linking skip the mention extraction step and assume the gold mentions in the datasets are given. These work only focus on improving the later stages of the entity linking pipeline.

4.1 English NER

Named entity recognition (NER), sometimes also called named entity recognition and classification (NERC), is the task of locating and typing named entity phrases in text. Below is an example of phrases annotated with named entities:

[U.N. *ORG*] official [Rolf Ekeus *PER*] heads for [Baghdad *LOC*].

In this example, brackets indicate mention boundaries and the last words inside brackets indicate named entity types. ORG, PER, and LOC refer to person, organizations, and locations respectively. The target named entity types and definition of types may vary across datasets. The above three entity types are the most common ones in the literature.

One difficulty of NER is that names form an open class of words. Although certain names are common and frequently appear in text, new names have been invented everyday. It is unrealistic to assume that one can collect a name list that covers all possible named entities. For example, organization names such as Slack and SpaceX are only recently coined. In English, and many other Latin- or Cyrillic-script languages, entities are marked with capitalization, which makes detection substantially easier. However, not all kind of text is well-formatted and capitalization can be used in several other situations. Therefore, systems must rely on context of the text in order to detect named entities reliably.

Similar to entity linking, the typing aspect of NER performs coarse-grained disambiguation. For example, NER could distinguish "Washington" as a person or a location, but several locations and people could be referred to as Washington. When using NER as a mention detector in the EDL pipeline, the predicted entity types are usually not needed, hence several EDL systems discard the entity types and only take the mention boundaries. However, some works find that incorporating NER types into an EDL model is helpful.

4.1 English NER

NER is considered a fundamental task for language understanding since named entities usually carry main information in an article. NER is proven to be useful for many NLP tasks besides EDL. For example, in machine translation, names are usually treated differently then other words since names can be very ambiguous. In question answering, answers to the questions are often named entities. In information extraction tasks such as relation extraction, event extraction, and semantic role labelling, many arguments are named entities.

4.1.1 Problem Formulation

Before discussing details of NER models, it is important to understand how NER is formulated as a machine learning problem.

NER is typically formulated as a sequence labeling task, in which each sentence is a sequence of words, and each word is assigned a tag. Since named entities can be multi-word phrases, these word-level tags should encode which phrase this word belongs to. There are a number of schemes for encoding phrases in sequential tags. The most common ones are IOB2 and BILOU labeling scheme. Using the earlier example with three named entity mentions,

[U.N. *ORG*] official [Rolf Ekeus *PER*] heads for [Baghdad *LOC*],

the labeled sentence using each scheme is:

- IOB2: [U.N. *B-ORG*] [official *O*] [Rolf *B-PER*] [Ekeus *I-PER*] [heads *O*] [for *O*] [Baghdad *B-LOC*]
- BILOU: [U.N. *U-ORG*] [official *O*] [Rolf *B-PER*] [Ekeus *L-PER*] [heads *O*] [for *O*] [Baghdad *U-LOC*]

In the IOB2 scheme, "B-" or "I-" is prepended to the entity types. "B-" (for Begin) indicates the beginning of a mention, whereas "I-" (for Inside) is used for all other words in entity mentions. Words do not belong to any entity mention are labeled as "O" (for Outside). Similar to IOB2, the BILOU scheme introduces two more prefixes, "U-" and "L-". Mentions with only one token take "U-", and the last token of a multi-token phrase takes "L-".

All schemes are technically equivalent and can be converted to each other deterministically without information loss. However, the numbers of unique labels are different. To calculate the number of labels:

number of labels = (number of non-O prefixes) × (number of entity types) + 1,

where the +1 in the end is for the "O" tag. For example, if there are three entity types, {*PER, LOG, ORG*}, the number of unique labels for IOB2 and BILOU schemes are 7 and 13 respectively.

The idea behind BILOU scheme is that words at the mention boundary should be treated differently since their neighboring words would be different from the ones inside an entity. For instance, the last word of a multi-word entity would have a name token to the left and likely a non-name token to the right. Therefore, the BILOU scheme is more expressive than the IOB2 scheme. The trade-off here is that using more labels might require more training data for machine learning models. Ratinov and Roth (2009) show that using the BILOU scheme yields better performance. However, Reimers and Gurevych (2017) suggest IOB2 outperforms BILOU. This discrepancy in conclusion could be due to the NER models studied in the two papers. Ratinov and Roth (2009) use a linear classifier with hand-crafted features, whereas Reimers and Gurevych (2017) study neural network based models. We will discuss these two types of models next.

4.1.2 Models

When building an NER system, one typically trains a machine learning model on some large corpus of text annotated with named entities. In this supervised learning setting, the space of NER models can be divided into earlier non-neural models and more recent neural architectures. We defer a comprehensive literature review to other NER-focused publications (Nadeau and Sekine 2007; Yadav and Bethard 2018; Li et al. 2020b), but give a brief overview of popular models.

4.1.2.1 Non-neural Models

Given any labeling scheme discussed in the previous section, NER is essentially a word classification problem. The models built before the neural age used features drawn from the current word and the neighboring words within a fixed size context window. The exact features would vary by implementation. We categorize features into two types based on what kinds of resources are used: **local features**, which only depend on the current sentence, and **external features**, which rely on other resources beyond the current text at hand. Here we list commonly used features:

- Local features:
  - Surface form of word
  - Word shape (e.g., capitalization pattern such as Aaaa, AAAA, or AaAa)
  - Word type (e.g., whether or not all characters are digits or non-alphabetic letters)
  - Affixes (the first and last k characters of a word, where k is usually from 3 to 5 for English)
  - Whether or not a word is capitalized
  - Whether or not the current word is the first word of a sentence
  - Part of speech

- External features:
  - Brown cluster bit strings, and prefixes of bit strings with various lengths
  - Gazetteer features. Whether or not the current word together with neighboring words are part of some name in pre-defined entity lists.

Brown clustering is a form of hierarchical clustering of words based on the contexts in which they occur. The result is essentially a binary tree where each leaf is a word. A word can be represented by the path from the root node to the corresponding leaf. Two semantically similar words would have longer common prefix in their representations. Details of the brown clustering algorithms can be found in Appendix A.1.1.

Note that besides extracting features from the current word, most of the features are also applied on the neighboring words. For example, there is a feature indicates the surface form of the previous word, and another feature indicates the word shape of the word before the previous word. Usually, including 2 words before and after of the current word is sufficient for English.

Several classification models have been used with these features, including Support Vector Machines (SVMs) (McNamee and Mayfield 2002), decision trees with AdaBoost (Carreras et al. 2002), averaged perceptron (Ratinov and Roth 2009), Hidden Markov Model (HMM) (Burger et al. 2002), and Conditional Random Fields (CRF) (McCallum and Li 2003; Finkel et al. 2005). For the first three classifiers (SVMs, decision trees, and averaged perceptron), each word is treated as an independent classification instance. That is, these classifiers predict a word's label solely use features extracted for this word. This independence assumption could have issues of producing impossible labeling sequence such as "B-PER I-ORG". This is because when the model predicts the label for the second word, it knows nothing about the label of the previous word. To address this issue, previous words' labels will be used as features. This would allow the model to learn that "I-ORG" is less likely to follow a "B-PER" token for example. In contrary, structured prediction models such as HMM and CRF do not suffer from this problem since they assign labels for all words in a sentence simultaneously. Considering all possible labelling sequences is intractable for long sentences as the number of label combinations grow exponentially with the length of the sentence. However, by assuming a first-order Markovian dependence between the labels, this can be computed efficiently with dynamic programming techniques because of the linear-chain structure. Since linear-chain CRF is widely used in various NER models to assign final label sequence, we give more detailed discussion in Appendix A.3.4.

There are two important pieces of software for non-neural NER models: CogCompNLP[1] (Ratinov and Roth 2009; Khashabi et al. 2018) and Stanford NER[2] (Manning et al. 2014). Both systems are written in Java, and are relatively easy to download and use, leading to widespread adoption.

---

[1] https://github.com/CogComp/cogcomp-nlp.
[2] https://nlp.stanford.edu/software/CRF-NER.shtml.

4.1.2.2 Neural Models

Deep learning models became popular for NER around the year of 2016. Comparing to the traditional models, neural methods do not require domain expertise for designing features. Instead, researchers can focus on designing model architectures for learning useful representations and underlying factors automatically. We follow the taxonomy proposed by Li et al. (2020b) and show the general model architecture in Fig. 4.2. The input sentence is first represented by some pre-trained word embeddings. A context encoder will run through the entire sentence and generate a vector for each word. The new representations from a context encoder carry more contextual information. Finally, similar to non-neural models, a label classifier is applied in the end to determine entity labels for all input words. In the following, we survey different choices of each of these components. Details about word representation and context representation models will be discussed in Appendices A.1 and A.2 respectively.

Distributed Word Representations The inputs to the model are usually distributed word representations (word embeddings), which are low dimensional real-valued dense vectors where each dimension represents a latent feature of the word. These distributed word representations are typically learnt from large collections of text. Some earlier well-known models are continuous bag-of-words (CBOW), skip-gram models Mikolov et al. (2013a, c), and GloVe Pennington et al. (2014).

Instead of only considering word-level information, character-level information has been shown to be important for NER such as the affixes features we saw in the section of non-neural models. Another advantage of character-level representation is that it could handle out-of-vocabulary words if these unseen words share some character-level regularity with

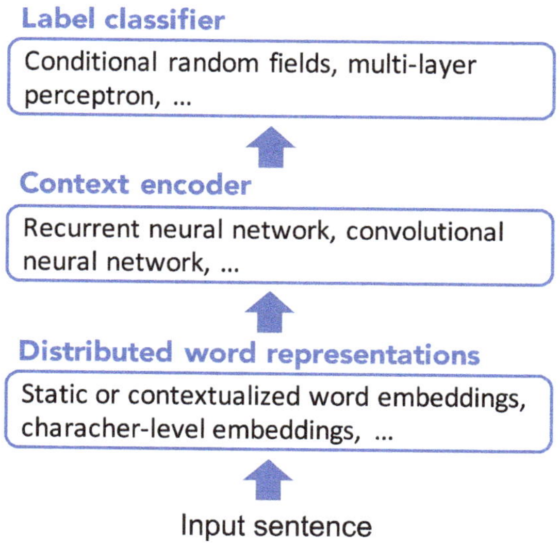

Fig. 4.2 A common model architecture for neural network based NER models

seen words. Ma and Hovy (2016) used a CNN for extracting character-level representations which are then concatenated with the word-level representations before feeding into a context encoder. Instead of using a CNN, Lample et al. (2016) used a bidirectional LSTM to extract character embeddings.

In addition to the aforementioned static word embeddings (a word always get the same vector regardless its context), contextualized word embeddings are shown to be superior in various NLP tasks. Contextualized word embedding models are pre-trained on large collection of text with some language modelling objective. At inference time, instead of looking up a pre-trained word vector, the pre-trained model is applied on the entire sentence to generate representations for each token in the sentence. In other words, the representation of a word depends on other words in the sentence. Peters et al. (2018) proposed ELMo word representations which are computed based on two-layer bidirectional language models (biLM) with character convolutions. The final word representation is a linear combination of representations from different layers of biLMs. Devlin et al. (2019) proposed BERT, bidirectional encoder representations from Transformers. BERT uses masked language model to enable deep bi-directional representations. Instead of using character representations, BERT pre-processes and splits words into WordPieces. Each input WordPiece token is represented by the sum of its token embedding, segment embedding, and position embedding. These input representations are then fed into several layers of Transformer encoders. Akbik et al. (2018) used character-level neural language models to generate a contextual word representation for a string of characters. They trained a forward and a backward character language models using LSTM. Their proposed word embeddings have been shown effective for NER task.

Context Encoder The input word representations are then passed through a context encoder which runs through the entire sentence and outputs a vector for each word. The most common context encoders are RNN or CNN. Recurrent neural networks have been shown to be effective in modeling sequential data. In particular, bidirectional RNNs are widely used as a context encoder for NER task. The forward RNN captures past information, whereas the backward RNN effectively uses future information. The representation from the forward and backward RNNs are then concatenated to form the output for a word. Huang et al. (2015) were among the first to apply bidirectionl LSTM (BiLSTM) as the context encoder for sequence tagging (including POS, chuncking, and NER). Later, BiLSTM-CRF architecture becomes the most popular model architecture for neural NER Lample et al. (2016), Ma and Hovy (2016).

When contextualized word embeddings are used as the input distributed word representations, the sentential information has been already encoded, therefore it might not necessary to apply another context encoder on top of them. For example, BERT Devlin et al. (2019) does not further add an NER-specific context encoder on top of several layers of Transformers. However, ELMo Peters et al. (2018) and Flair Akbik et al. (2018) still apply a BiLSTM encoder on top of their their contextualized word representations.

Label Classifier The final layer of the neural architecture is a label decoder, which takes the token representations produced by the context encoder and outputs a label (e.g., B-LOC,

I-ORG, etc.) for each token. This is similar to the classification models used in non-neural models. The difference is that for non-neural models, the input features to a classifier are usually hand-crafted according to some domain knowledge, whereas the features in neural models are generated by some neural networks, and thus it is usually hard to understand the meaning behind these features.

Given the success of Conditional Random Fields (CRF) in non-neural NER models, several deep learning based NER models also use CRF as the label decoder (Huang et al. 2015; Ma and Hovy 2016; Lample et al. 2016). Other structured prediction models such as structured Support Vector Machines has also been adopted as the label decoder in a neural NER model (Arora et al. 2019). A draw back of CRF or the common sequence tagging models is that only word-level information is used. Namely, words are treated as the base units. Semi-Markov CRF tries to overcome this issue by considering all possible segmentation (with some maximum segment length) of a sentence, so one could add segment-level or phrase-level features from multiple words together. Zhuo et al. (2016), Ye and Ling (2018) try to bring in semi-Markov CRF in neural NER models.

Besides the more expressive structured prediction models, several works simply use a local model as the label decoder. As we discussed in the non-neural model section, a local model does not assign labels for all words in the sentence simultaneously. Instead, it makes prediction on a word only based on the features at this token position. For neural NER models, a common choice of local label decoder is multi-layer perceptrons (Strubell et al. 2017; Devlin et al. 2019).

4.1.3 English NER Datasets

High quality annotations are crucial for both model learning and reliable evaluation. The most widely used English NER dataset is CoNLL 2003 (Tjong Kim Sang and De Meulder 2003), which was created for a shared task on language independent NER for the Conference on Natural Language Learning (CoNLL) in 2003. This corpus consists of Reuters news stories between August 1996 and August 1997. The target four named entity types are persons (PER), locations (LOC), organizations (ORG), and miscellaneous names (MISC). This MISC type is a broadly-defined catch-all that contains entities such as event, language, nationality, and product. CoNLL 2003 has been the primary benchmark for English NER ever since its creation.

Another popular NER dataset is OntoNotes (Hovy et al. 2006),[3] with a rich set of annotations for English, Arabic, and Chinese. The NER dataset contains 18 entity types. Besides PER, LOC, and ORG, some examples of additional types are NORP (nationalities), FACILITY (buildings, airports, etc.), GPE (geopolitical entities such as countries and cities), PRODUCT (vehicles, weapons, etc.), EVENT (battles, wars, sports events, etc.), and WORK

---

[3] The latest version at the time of writing is 5.0 (Weischedel et al. 2013).

OF ART (books, songs, etc.). Values are also annotated with tags such as TIME, DATE, and MONEY. OntoNotes dataset is an order of magnitude larger than CoNLL 2003 dataset, and covers various genres include newswire, broadcast news, broadcast conversation, telephone conversation, and web data.

Some years of the TAC KBP entity linking data can be used to train and evaluate NER as well. Since TAC 2014, English named entity mentions are annotated exhaustively with the corresponding entity types (Table 2.2). The focused named entity types in TAC datasets are PER, ORG, LOC, and GPE.

We only pointed out key datasets in this section. Besides these in the general domain, there are also several datasets in specialized domains such as twitter (Derczynski et al. 2016, 2017), code-switched data, or biomedical applications (Islamaj Doğan and Lu 2012). Interested readers could refer to NER survey papers such as Li et al. (2020b) for a more complete list of datasets.

4.1.4 English NER for EDL

NER models has been used as the mention extractor for EDL since knowledge bases consist of mostly named entities. For instance, one of the earliest Wikification work, Cucerzan (2007), truecases the input text and then applies an named entity recognizer to extract mentions. In addition, many entity linking works were driven by several years of the Text Analysis Conference Knowledge Base Population shared tasks in which the targeted mentions are named entities.

In addition to the common pipeline approach that fully trust NER outputs, several researchers have investigated models which jointly solve NER and Entity Linking since the two tasks may reinforce each other. On one hand, NER models usually struggle with long mentions (with several words). For instance, for longer mentions such as "The New York Times" or "Romeo and Juliet", NER models tend to extract partial names: "New York", "Romeo", and "Juliet". An Entity Linking model may be able to help to connect Romeo and Juliet if "Romeo and Juliet" is an plausible entity in the context. On the other hand, the coarse named entity type of a mention may help to restrict the space of entity candidate, therefore simplifies the linking problem. For instance, if a mention "Washington" is tagged as a LOCATION by an NER model, the entity linking model can then pay more attention to the location entities instead of people who can be referred to as Washington as well.

Over-generating Mention Candidates A common idea here is to over-generate mention candidates and apply an Entity Linking model to find the target entities for every mentions. The mention candidates with low confidence scores will then be rejected, therefore will not be included in the final answers. We discuss a couple of papers and highlight they key ideas which fall into this category.

Guo et al. (2013) formulate mention detection and disambiguation together as a structured prediction problem. Each *n*-gram that matches some anchor text becomes a mention candidate. The structural SVM model jointly grounds mention candidates to Wikipedia or rejects mention candidates. Since they focus on twitter data, the proposed joint learning and inference framework is feasible since tweets are very short.

Similar to this idea, Sil and Yates (2013) also over-generates mention candidates and let the model to link or to reject the mention candidates simultaneously. However, since news articles are much longer than tweets, it is not tractable to consider all n-grams as mention candidates. Instead, they use the mentions detected by an NER, the noun phrases identified by a shallow parser, and some heuristic rules to expand the existing mention candidates. Furthermore, they partition mention candidates in a document into several groups based on how close they are to each other. The model only performs joint prediction on the mentions in the same group.

A more recent work, Kolitsas et al. (2018), also follow the same paradigm. They focus on general news documents and use a neural network model. Given a document, all possibly linkable token spans are considered as mention candidates. Namely, all n-grams that have at least one entity candidate from Wikipedia. If the top entity candidate has a model score less than a threshold, this mention candidate will be rejected, therefore it will not be extracted as a mention. The large number of mention candidates make it difficult to perform joint inference on multiple mentions together. One could simply disambiguate each mention candidate independently. The authors also propose an approach to carefully select which mention candidates could participate in their global disambiguation model. We will discuss more about the local and global inference approaches for Entity Linking in Chap. 6.

Entity as Part of NER Labels Another idea is to incorporate entity labels into sequence tagging models. For instance, in Nguyen et al. (2016), the label for each token becomes a concatenation of the NER label and Wikipedia title (e.g., PER:Barack_Obama). Consecutive tokens with identical labels are consider entity mentions. For instance, only if we have "Barack [PER:Barack_Obama] Obama [PER:Barack_Obama]", "Barack Obama" will be identified as a mention. Instead of the widely used linear-chain model, the authors propose a tree model in which the factor connections between tokens are based on the results of a dependency parser. In their experiment, the tree model outperforms the linear chain model on both NER and end-to-end entity linking tasks. Chen et al. (2020) propose a similar idea using a BERT-base model. At inference time, the model assigns a mention boundary label (B , I, or O) and an entity label (entity ID in the KB) to each token. The entity ID of the first token of a mention (with a B tag) will be used as the linked entity for this mention.

As we can see in the previous two works, if Entity Linking is performed on token level, a multi-token mention may be linked to several entities. Therefore some rules need to be applied to resolve this discrepancy. This issue can be addressed by using a semi-Markov CRF (Sarawagi and Cohen 2004) which considers all possible segmentations of a sequence so that labels are assigned to segments instead of words. Luo et al. (2015) also jointly model NER and entity linking by incorporating ranking features for disambiguation into an NER model. They extend semi-Markov CRF to not only directly models mention boundaries but also considers entity distribution and mutual dependency over segmentations.

4.2 NER in Other Languages

Most cross-lingual EDL work focuses on disambiguating named entities written in a non-English language. Extracting non-English named entity mentions becomes the first task in the cross-lingual EDL pipeline. A key challenge of non-English NER is that there is lack of training data in the target language, since a good performing machine learning model usually learns from lots of examples. As discussed in the previous sections, there are several high-quality corpora for English NER, and the state-of-the-art English NER models rely on such resources. However, there is no such rich resource for most non-English languages. To overcome this obstacle, researchers have developed techniques and models for either creating target language annotations automatically by leveraging resources such as Wikipedia and parallel corpora, or transferring an English NER model to the target language by cross-lingual language representations.

When high quality target-language training datasets are available, the architecture of the target-language NER model would be very similar to the English models that we discussed in the previous sections. In this section, we will focus on the low-resource scenario where there is no target-language NER training data and therefore some cross-lingual transfer techniques are needed. We organize this section by what kinds of resources are leveraged to cross the language barrier, ordered roughly from most expensive to least expensive.

4.2.1 Human Annotation

To build a target-language NER model, the most direct way is to annotate named entity mentions in some target-language documents, and then train a monolingual NER model on these documents. Asking native speakers of the target language to perform annotation manually would give the best quality training data. This process is usually quite time-consuming and labor-intensive, but the resulting NER models usually perform better than other approaches if there is a good amount of training examples. Using this approach, one can directly obtain training examples in the domain of interest (e.g., text from news, social media, etc.), therefore there is no need to be worry about domain shift between training and test data.

The key non-English NER datasets includes Spanish and Dutch from CoNLL 2002 shared task (Tjong Kim Sang 2002), German from CoNLL 2003 shared task (Tjong Kim Sang and De Meulder 2003), and Chinese and Arabic from OntoNotes (Hovy et al. 2006). The models designed for these languages are largely similar to the English models that we discussed in this chapter in terms of both hand-crafted features and the neural architecture. However, considering language-specific properties could further improve the model performance. One example is explicitly handling morphology for the morphologically rich languages. For agglutinative languages such as Turkish and Arabic, a word could be inflated into a long word with complex morphological structure. Simply using word forms as features would produce several out-of-vocabulary words between training and test sets. Even using character-level embeddings may not fully encode the morphological structure. To overcome this issue, a common approach is to include some morphological analysis into the NER model. For instance, Yeniterzi (2011) designed features based on the outputs of a morphological disambiguator. For another example, Güngör et al. (2018) proposed a model which jointly learns NER and morphological disambiguation.

Although having human annotations is useful for training a good NER model, these high-quality corpora only exist for the languages which have lots of speakers. The requirement of finding native speakers of the target language who can perform the annotation job cannot always be satisfied easily, especially for low-resource languages. To overcome this issue, researchers (Mayhew and Roth 2018; Lin et al. 2018) have developed annotation platforms that help annotators who do not speak the target language to identify named entity mentions in documents. The key idea applied in these works is romanization, a process which turns any language script into Latin script while preserving the original word pronunciation. Since many named entities are transliterated into other languages, the pronunciation across different language could be pretty close. This property allows a non-speaker to be able to identify some entities. In addition, these annotation platforms also incorporate other useful tools such as translation by lexicons, TF-IDF token statistics, Internet search, and entity propagation.

Figure 4.3 shows a screenshot taken from TALEN (Mayhew and Roth 2018), a tool that helps non-speaker to annotate named entities in low-resource languages. In the document-based annotation screen, a romanized document from an Amharic corpus is shown. The user has selected "nagaso gidadane" (Negasso Gidada) for tagging, indicated by the thin gray border, and by the popover component. The lexicon (dictionary) is active, so the words which can be found in a Amharic-to-English dictionary will be displayed in the corresponding English translation. For instance, the "doctor" (in italics) immediately prior to the selected "nagaso gidadane" is translated from the original Amharic word at that position. We can see that this word could indicate that the following words represent a person. On the right panel, it shows the TF-IDF scores of the selected words, which are useful for preventing some obvious limitations such as labeling stop words as entities.

4.2.2 Parallel Text

Parallel text, in which sentences in one language are aligned with translations in another language, is often used to train machine translation models. One common idea of using parallel text for multilingual NLP tasks is to project annotations from one language to another language. Namely, labels of words in one language can be propagated to the words of another language via word alignments. Although concepts can be expressed in different ways across different languages, it is often reasonable to assume that word alignments calculated automatically can give word-level translations. For example, if we would like to

4.2 NER in Other Languages

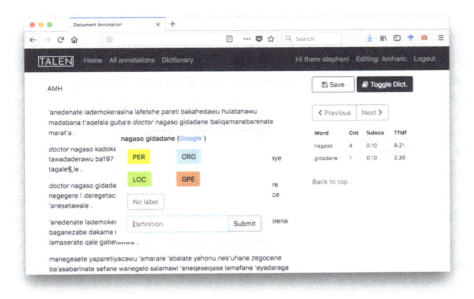

Fig. 4.3 A screenshot of TALEN (Mayhew and Roth 2018), a tool for annotating low-resource named entities

train a Spanish NER model, we can first take a parallel corpus between English and Spanish, and then annotate named entity mentions in the English documents by either humans or an English NER model. We can compute word alignments between parallel sentences and then annotate Spanish words using the labels of the corresponding English words. Finally, an NER model can be trained on the annotated Spanish documents in the monolingual fashion. This idea has been applied to several NLP tasks such as NP chunking (Yarowsky and Ngai 2001; Yarowsky et al. 2001), part-of-speech tagging (Das and Petrov 2011), NER (Ehrmann et al. 2011; Kim et al. 2012; Wang and Manning 2014), and parsing (Hwa et al. 2005; Zeman and Resnik 2008; Ganchev et al. 2009; McDonald et al. 2011).

When applying this projection technique, most earlier work does hard projection, in which labels are simply projected across alignments without any modification. Wang and Manning (2014) propose to project expectations over the labels instead. Namely, they first compute the posterior marginal at each word position on the English side using an English NER model; then for each aligned English word, the posterior marginal is projected to the aligned word position on the foreign side. They encode the expectations as constraints and train a model by minimizing divergence between model expectations and projected expectations. They show this soft projection approach improves over the hard projection variant on Chinese and German datasets. Similar idea were also found to be useful for part-of-speech tagging (Agić et al. 2015; Agić et al. 2016).

Another line of work that uses parallel text is to generate cross-lingual word representations. The idea is to embed words from different languages into the same vector space, so that words with similar meaning will be closer in that space regardless the language identity. Using these cross-lingual word representations as features or as inputs to a neural network model, one could train a model on English NER dataset, and directly apply the model on text written in another language. Täckström et al. (2012) used parallel text to induce cross-lingual word clusters, which are then used as features in a direct transfer model. The idea of using parallel text to produce cross-lingual word embeddings is also later pursued in Hermann and Blunsom (2014), Luong et al. (2015), and Lample and Conneau (2019).

Although parallel text provide strong relationship between two languages, it is quite hard to be obtained for many languages. Even when parallel text is available, there are different levels of quality and quantity. Moreover, the domain of parallel text could be different from the target-language text from which we would like to extract named entities. For example, Bible has been translated into many languages, but an NER model trained on Bible might not work well on other genres of text. Enghoff et al. (2018) study cross-lingual projection for NER on 17 languages, and suggest that it is not as simple as it seems because of the size and quality of parallel text.

4.2.3 Wikipedia

While high quality parallel text only exists for high-resource languages, Wikipedia is available in more than 300 languages. As introduced in Sect. 2.1.1, Wikipedia contains lots of human annotations and edits, which are not only useful for entity linking, but are valuable for NER. For example, the hyperlinked text could also be used as NER annotations if one can identify the entity type of the linked entity. That is, if we know the Wikipeida page "New York City" belongs to LOC (location) entity type, all the phrases that are linked to "New York City" can be tagged as LOC. One line of work (Nothman et al. 2008; Balasuriya et al. 2009; Nothman et al. 2009; Nothman et al. 2013) explores this idea and generates *silver standard* NER annotations on Wikipedia articles. This approach can be applied on any language in Wikipedia in principle, but several heuristics developed in earlier papers are language-specific rules. More recently, Pan et al. (2017) propose a classification model that classifies English Wikipedia pages into an NER tag set, and propagate annotations from English to other languages through the cross-language links in Wikipedia. Since not every named entity mention is hyperlinked in Wikipedia articles, they further apply the self-training technique to enrich and propagate annotations.

Besides creating silver-standard annotations on Wikipedia articles by leveraging anchor text and cross-language links, Kim et al. (2012) also generate parallel sentences from Wikipedia. The NER annotations on the English sentences then can be projected to foreign language sentences using the idea we discussed in the previous section of using parallel text.

4.2 NER in Other Languages

NER Tags:			Person			Location
Sentence:	Schwierigkeiten beim nachvollziehenden Verstehen Albrecht Lehmann läßt Flüchtlinge und Vertriebene in Westdeutschland					
Wikipedia titles:	Problem_solving	Understanding	Albert,_Duke_of_Prussia	Jens_Lehmann	Refugee	Western_Germany
FreeBase types:	hobby media_genre	media_common quotation_subject	person noble_person	person athlete	field_of_study literature_subject	location country

Fig. 4.4 An example of the language-independent features proposed by Tsai et al. (2016). Each word in the German sentence is grounded to English Wikipedia entries. The corresponding FreeBase types of these Wikipedia entries are used as NER features for any language that is supported by the cross-lingual entity linking model

Wikipedia has also been used to generate cross-lingual word representations. Similar to Täckström et al. (2012), but instead of using parallel text, Tsai et al. (2016) propose language-independent NER features based on a cross-lingual entity linking model. The key idea is to represent words in any language by Freebase types and Wikipedia categories. This is achieved by linking every n-grams in the foreign text to the English Wikipedia.[4] Figure 4.4 shows an example of German sentence. The most probable Wikipedia entry of each word returned by a cross-lingual entity linking model is listed. We can see that even though the disambiguation is not perfect, the FreeBase types still provide valuable information for recognizing which words belong to named entities. Although the person "Albrecht Lehmann" does not exist in Wikipedia, the model still links both "Albrecht" and "Lehmann" to some other people, therefore the FreeBase type "person" is one of the features for these two words. Since these features are always in English, they are language-independent, which allows a trained NER model to be applied on any language directly.

Rijhwani et al. (2020) also ground every n-grams to the English Wikipedia. But instead of using the typing information as explicit features, they use the scores obtained from the cross-lingual entity linking model as "soft gazetteer" features. These features are incorporated into a neural NER model via a jointly trained autoencoder objective.

4.2.4 Large Multilingual Pre-trained Language Models

In the previous two sections, we have seen researchers tried to generate multilingual representation of words using parallel text or Wikipedia. With multilingual word representations, a model trained on one language could be applied to another language directly. In recent years, massively pre-trained English language models have been shown to provide good word or sentence embeddings, therefore can achieve state-of-the-art results for many NLP tasks. These methods often make use of the transfer learning paradigm—language models are pre-trained on a large amount of plain text, and then fine-tuned on a downstream task of

---

[4] We will discuss details of cross-lingual entity linking models in the next two chapters.

interest with task-specific training data. Some examples of these models are BERT (Devlin et al. 2019), RoBERTa (Liu et al. 2019), BART (Lewis et al. 2020), and T5 (Raffel et al. 2020).

Most of these successful models have been extended to generate multilingual representations. For instance, mBERT (Devlin et al. 2019)[5] is a multilingual version of BERT. Instead of training on the English Wikipedia and the Toronto Books Corpus, mBERT is trained on the top 100 languages from Wikipedia. XLM-R (Conneau et al. 2020) is based on the RoBERTa model. It is trained with a cross-lingual masked language model on 100 languages from the Common Crawl data. mBART (Liu et al. 2020) extended BART to become multilingual. Last but not least, mT5 (Xue et al. 2021) is a multilingual version of the T5 model, which is also trained on multilingual documents from the Common Crawl corpus.

These multilingual language models are different in a couple of aspects. For training data, the most two common data sources are multilingual Wikipedia and Common Crawl. Different models have applied different strategies in cleaning, selecting, and pre-processing sentences from a massive amount of text. This is especially important when using Comomon Crawl data. For example, identifying the language of a document is not always perfectly accurate. Data source could also impact the number of supported languages for a model. Regarding modeling, most models are built upon the Transformer architecture. However, models such as mBERT and XLM-R only use the encoder part, whereas mBART and mT5 follow the encoder-decoder architecture.

These models are usually evaluated on several cross-lingual NLP tasks including cross-lingual NER. XTREME (Hu et al. 2020) is a widely used multilingual multi-task benchmark which contains 9 tasks across 40 languages. The multilingual NER task in XTREME is based on Wikiann (Pan et al. 2017) dataset, in which named entities in Wikipedia are annotated with an NER tag set automatically using heuristics and self-training techniques. When evaluating these multilingual language models on NER, there are two common settings: only fine-tuning on English NER annoatations or fine-tuning on the target language NER annotations. The former essentially tests models' cross-lingual zero-shot transfer ability, whereas the latter is an in-language multitask setup.

4.2.5 Lexicons

Although Wikipedia is a useful multilingual resource and it covers more than 300 languages, the size (number of articles) of different languages' Wikipedia varies. The usefulness of Wikipedia decreases with the size of Wikipedia since a small Wikipedia contains fewer human annotations (hyperlinks, inter-language links, etc.), which are the key signals for cross-lingual NER and entity linking.'

---

[5] Appendix A.2.2.

4.2 NER in Other Languages

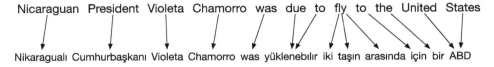

Fig. 4.5 An word translation example taken from Mayhew et al. (2017). The top is English, and the bottom is Turkish. Arrows represent dictionary translations. **Correct** is the correct translation. This example illustrates congruence in named entity patterns between languages

For those languages that lack parallel text and large Wikipedia, it is often possible to find lexicons or dictionaries, which contain word-level mapping between two languages. For instance, PanLex (Kamholz et al. 2014) project covers lexicons for more than 5,000 languages. Although these word translations are without contextual information, they are proven to provide good cross-lingual signals in many NLP applications.

One line of work uses lexicons to create cross-lingual language representations. Similar to the ones created using parallel text or Wikipedia. If words in different languages are mapped into the same semantic space, one can direct transfer an NER model trained on some high-resource language. One representative work along this line is Faruqui and Dyer (2014), in which they propose to map monolingual word embeddings of two languages into the same space using Canonical Correlation Analysis. The word embeddings from two languages are aligned according to a lexicon. These aligned pairs are then used to compute the transformation matrices which map monolingual embeddings into the new cross-lingual vector space. We will discuss more about cross-lingual word representations in Appendix A.1.4.

Besides creating cross-lingual word representations, Mayhew et al. (2017) make use of a lexicon to "translate" annotated data available in one or several high-resource language(s) into the target language, and learn a monolingual NER model on the projected labels. Figure 4.5 shows an example of word translation between English and Turkish. We see that some words are translated correctly (e.g., "President" into "Cumhurbaçkanı") and some incorrectly (e.g., "fly" into "iki taçın arasıda"). There is also ignorance of morphology, and confused word order. In spite of all these mistakes, the context around the entities, which is the key signal for NER, is reasonably well-preserved. They show that this simple *cheap translation* approach gives non-trivial scores across several languages in the cross-lingual NER setting.

4.3 Bibliographical Notes

Although NER is commonly used as the mention extractor in the EDL pipeline, several earlier works on Entity Linking extract general key phrases instead. For instance, Mihalcea and Csomai (2007) view mention extraction as a keyword extraction problem, and propose an unsupervised approach. They construct a controlled vocabulary which contains all Wikipedia titles and frequent anchor texts. Given a piece of text, they first extract all n-grams which match some entries in the vocabulary. The matched n-gram candidates are then ranked by their "keyphraseness", which measures how often a term is hyperlinked in the entire collection of Wikipedia documents.

Milne and Witten (2008) learn a supervised classifier from Wikipedia anchor text to decide if an n-gram is a mention. Similar to the idea of jointly modeling NER and Entity Linking, they use the later steps of the pipeline to disambiguate every n-grams in order to generate better features for this mention classifier. Some of the features they use indicate the generality, location, and spread of the phrase. These features try to capture the ideas that the phrases mentioned in the first paragraph tend to be more important, and that how consistently the document discusses this phrase.

Ling et al. (2015) discuss different annotation styles among EDL datasets. Some dataset may include mentions of general concepts in addition to named entities. Even in the datasets contain only named entity mentions, the types of the entities may vary. For instance, most years of TAC-KBP tasks focus on four entity types: Person, Organization, Location, and Geo-political entities, whereas the most popular NER benchmark, CoNLL dataset, includes Person, Organization, Location, and Misc.

For English NER, there are also several datasets in the biomedical domain. Some examples are Islamaj Doğan and Lu (2012), BC5CDR,[6] FSU-PRGE,[7] and GENETAG.[8] More datasets and comparison between them can be found in Crichton et al. (2017).

Besides the works cited in Sect. 4.2.1 that utilize language specific properties, there is a line of research on NER for morphologically rich languages (Demir and Özgür 2014; Şeker and Eryiğit 2012; Tür 2000; Hasan et al. 2009). There are also many studies on monolingual NER models for non-English high-resource languages. For instance, Chinese NER could be more challenging than English NER since word segmentation or tokenization is more difficult in Chinese. Several SIGHAN shared tasks (Zhang et al. 2006; Jin and Chen 2008; Mao et al. 2008; He et al. 2012) have focused on Chinese NER in news domain. Later, Nie et al. (2020), Peng and Dredze (2016, 2015) propose NER models for Chinese social media text. More recently, various types of deep learning models (Zhang and Yang 2018; Zhu and Wang 2019; Gui et al. 2019) have been developed to carefully use both character and word level information in order to alleviate word segmentation errors.

---

[6] http://bioc.sourceforge.net/.
[7] https://julielab.de/Resources/FSU_PRGE.html.
[8] https://julielab.de/Resources/FSU_PRGE.html.

References

Agić, Ž., Hovy, D., Søgaard, A.: If all you have is a bit of the Bible: Learning POS taggers for truly low-resource languages. In: Proceedings of the 53rd Annual Meeting of the Association for Computational Linguistics and the 7th International Joint Conference on Natural Language Processing (Volume 2: Short Papers), Association for Computational Linguistics, pp. 268–272. Beijing, China (2015). https://doi.org/10.3115/v1/P15-2044, https://aclanthology.org/P15-2044

Agić, Ž., Johannsen, A., Plank, B., Martínez Alonso, H., Schluter, N., Søgaard, A.: Multilingual projection for parsing truly low-resource languages. Trans. Assoc. Comput. Linguistics **4**, 301–312 (2016). https://doi.org/10.1162/tacl_a_00100, https://aclanthology.org/Q16-1022

Akbik, A., Blythe, D., Vollgraf, R.: Contextual string embeddings for sequence labeling. In: Proceedings of the 27th International Conference on Computational Linguistics, Association for Computational Linguistics, pp. 1638–1649, Santa Fe, New Mexico, USA (2018). https://aclanthology.org/C18-1139

Arora, R., Tsai, C.-T., Tsereteli, K., Kambadur, P., Yang, Y.: A semi-Markov structured support vector machine model for high-precision named entity recognition. In: Proceedings of the 57th Annual Meeting of the Association for Computational Linguistics, Association for Computational Linguistic, pp. 5862–5866. Florence, Italy (2019). https://doi.org/10.18653/v1/P19-1587, https://aclanthology.org/P19-1587

Balasuriya, D., Ringland, N., Nothman, J., Murphy, T., Curran, J.R.: Named entity recognition in Wikipedia. In: Proceedings of the 2009 Workshop on The People's Web Meets NLP: Collaboratively Constructed Semantic Resources (People's Web), Association for Computational Linguistics, pp. 10–18. Suntec, Singapore (2009). https://aclanthology.org/W09-3302

Burger, J.D., Henderson, J.C., Morgan, W.T.: Statistical named entity recognizer adaptation. In: COLING-02: The 6th Conference on Natural Language Learning 2002 (CoNLL-2002) (2002). https://aclanthology.org/W02-2003

Carreras, X., Màrquez, L., Padró, L.: Named entity extraction using AdaBoost. In: COLING-02: The 6th Conference on Natural Language Learning 2002 (CoNLL-2002) (2002). https://aclanthology.org/W02-2004

Chen, H., Li, X., Zukov Gregoric, A., Wadhwa, S.: Contextualized end-to-end neural entity linking. In: Proceedings of the 1st Conference of the Asia-Pacific Chapter of the Association for Computational Linguistics and the 10th International Joint Conference on Natural Language Processing, Association for Computational Linguistics, pp. 637–642. Suzhou, China (2020). https://aclanthology.org/2020.aacl-main.64

Conneau, A., Khandelwal, K., Goyal, N., Chaudhary, V., Wenzek, G., Guzmán, F., Grave, E., Ott, M., Zettlemoyer, L., Stoyanov, V.: Unsupervised cross-lingual representation learning at scale. In: Proceedings of the 58th Annual Meeting of the Association for Computational Linguistics, Association for Computational Linguistics, pp. 8440–8451, Online (2020). https://doi.org/10.18653/v1/2020.acl-main.747, https://aclanthology.org/2020.acl-main.747

Crichton, G., Pyysalo, S., Chiu, B., Korhonen, A.: A neural network multi-task learning approach to biomedical named entity recognition. BMC Bioinform. **18**(1), 368 (2017)

Cucerzan, S.: Large-scale named entity disambiguation based on Wikipedia data. In: Proceedings of the 2007 Joint Conference on Empirical Methods in Natural Language Processing and Computational Natural Language Learning (EMNLP-CoNLL), Association for Computational Linguistics, pp. 708–716, Prague, Czech Republic (2007). https://aclanthology.org/D07-1074

Das, D., Petrov, S.: Unsupervised part-of-speech tagging with bilingual graph-based projections. In: Proceedings of the 49th Annual Meeting of the Association for Computational Linguistics: Human Language Technologies, Association for Computational Linguistics, pp. 600–609. Portland, Oregon, USA (2011). https://aclanthology.org/P11-1061

Demir, H., Özgür, A.: Improving named entity recognition for morphologically rich languages using word embeddings. In: 2014 13th International Conference on Machine Learning and Applications, pp. 117–122. IEEE (2014)

Derczynski, L., Bontcheva, K., Roberts, I.: Broad Twitter corpus: a diverse named entity recognition resource. In: Proceedings of COLING 2016, the 26th International Conference on Computational Linguistics: Technical Papers, The COLING 2016 Organizing Committee, pp. 1169–1179. Osaka, Japan (2016). https://aclanthology.org/C16-1111

Derczynski, L., Nichols, E., van Erp, M., Limsopatham, N.: Results of the WNUT2017 shared task on novel and emerging entity recognition. In: Proceedings of the 3rd Workshop on Noisy User-generated Text, pp. 140–147, Association for Computational Linguistics. Copenhagen, Denmark (2017). https://doi.org/10.18653/v1/W17-4418, https://aclanthology.org/W17-4418

Devlin, J., Chang, M.-W., Lee, K., Toutanova, K.: BERT: pre-training of deep bidirectional transformers for language understanding. In: Proceedings of the 2019 Conference of the North American Chapter of the Association for Computational Linguistics: Human Language Technologies, Volume 1 (Long and Short Papers), Association for Computational Linguistics, pp. 4171–4186. Minneapolis, Minnesota (2019). https://doi.org/10.18653/v1/N19-1423, https://aclanthology.org/N19-1423

Ehrmann, M., Turchi, M., Steinberger, R.: Building a multilingual named entity-annotated corpus using annotation projection. In: Proceedings of the International Conference Recent Advances in Natural Language Processing 2011, Association for Computational Linguistics, pp. 118–124. Hissar, Bulgaria (2011). https://aclanthology.org/R11-1017

Enghoff, J.V., Harrison, S., Agić, Ž.: Low-resource named entity recognition via multi-source projection: Not quite there yet? In: Proceedings of the 2018 EMNLP Workshop W-NUT: The 4th Workshop on Noisy User-generated Text, Association for Computational Linguistics, pp. 195–201. Brussels, Belgium (2018). https://doi.org/10.18653/v1/W18-6125, https://aclanthology.org/W18-6125

Faruqui, M., Dyer, C.: Improving vector space word representations using multilingual correlation. In: Proceedings of the 14th Conference of the European Chapter of the Association for Computational Linguistics, pp. 462–471. Gothenburg, Sweden (2014). https://doi.org/10.3115/v1/E14-1049, https://aclanthology.org/E14-1049

Finkel, J.R., Grenager, T., Manning, C.: Incorporating non-local information into information extraction systems by Gibbs sampling. In: Proceedings of the 43rd Annual Meeting of the Association for Computational Linguistics (ACL'05), Association for Computational Linguistics, pp. 363–370. Ann Arbor, Michigan (2005). https://doi.org/10.3115/1219840.1219885, https://aclanthology.org/P05-1045

Ganchev, K., Gillenwater, J., Taskar, B.: Dependency grammar induction via bitext projection constraints. In: Proceedings of the Joint Conference of the 47th Annual Meeting of the ACL and the 4th International Joint Conference on Natural Language Processing of the AFNLP, Association for Computational Linguistics, pp. 369–377. Suntec, Singapore (2009). https://aclanthology.org/P09-1042

Gui, T., Zou, Y., Zhang, Q., Peng, M., Fu, J., Wei, Z., Huang, X.: A lexicon-based graph neural network for Chinese NER. In: Proceedings of the 2019 Conference on Empirical Methods in Natural Language Processing and the 9th International Joint Conference on Natural Language Processing (EMNLP-IJCNLP), Association for Computational Linguistics, pp. 1040–1050. Hong Kong, China (2019). https://doi.org/10.18653/v1/D19-1096, https://aclanthology.org/D19-1096

Güngör, O., Uskudarli, S., Güngör, T.: Improving named entity recognition by jointly learning to disambiguate morphological tags. In: Proceedings of the 27th International Conference on Computational Linguistics, Association for Computational Linguistics, pp. 2082–2092. Santa Fe, New Mexico, USA (2018). https://aclanthology.org/C18-1177

References

Guo, S., Chang, M.-W., Kiciman, E.: To link or not to link? a study on end-to-end tweet entity linking. In: Proceedings of the 2013 Conference of the North American Chapter of the Association for Computational Linguistics: Human Language Technologies, Association for Computational Linguistics, pp. 1020–1030. Atlanta, Georgia (2013). https://aclanthology.org/N13-1122

Hasan, K.S., Rahman, M.A.U., Ng, V.: Learning-based named entity recognition for morphologically-rich, resource-scarce languages. In: Proceedings of the 12th Conference of the European Chapter of the ACL (EACL 2009), Association for Computational Linguistics, pp. 354–362. Athens, Greece (2009). https://aclanthology.org/E09-1041

He, Z., Wang, H., Li, S.: The task 2 of CIPS-SIGHAN 2012 named entity recognition and disambiguation in Chinese bakeoff. In: Proceedings of the Second CIPS-SIGHAN Joint Conference on Chinese Language Processing, Association for Computational Linguistics, pp. 108–114. Tianjin, China (2012). https://aclanthology.org/W12-6321

Hermann, K.M., Blunsom, P.: Multilingual distributed representations without word alignment. In: Proceedings of the International Conference on Learning Representations (2014)

Hovy, E., Marcus, M., Palmer, M., Ramshaw, L., Weischedel, R.: OntoNotes: the 90% solution. In: Proceedings of the Human Language Technology Conference of the NAACL, Companion Volume: Short Papers, Association for Computational Linguistics, pp. 57–60, New York City, USA (2006). https://aclanthology.org/N06-2015

Hu, J., Ruder, S., Siddhant, A., Neubig, G., Firat, O., Johnson, M.: Xtreme: a massively multilingual multi-task benchmark for evaluating cross-lingual generalisation. In: International Conference on Machine Learning, pp. 4411–4421. PMLR (2020)

Huang, Z., Xu, W., Yu, K.: Bidirectional LSTM-CRF models for sequence tagging. arXiv preprint arXiv:1508.01991 (2015)

Hwa, R., Resnik, P., Weinberg, A., Cabezas, C.I., Kolak, O.: Bootstrapping parsers via syntactic projection across parallel texts. Nat. Language Eng. (2005)

Islamaj Doğan, R., Lu, Z.: An improved corpus of disease mentions in PubMed citations. In: BioNLP: Proceedings of the 2012 Workshop on Biomedical Natural Language Processing, Association for Computational Linguistics, pp. 91–99. Montréal, Canada (2012). https://aclanthology.org/W12-2411

Jin, G., Chen, X.: The fourth international Chinese language processing bakeoff: Chinese word segmentation, named entity recognition and Chinese POS tagging. In: Proceedings of the Sixth SIGHAN Workshop on Chinese Language Processing (2008). https://aclanthology.org/I08-4010

Kamholz, D., Pool, J., Colowick, S.: PanLex: building a resource for panlingual lexical translation. In: Proceedings of the Ninth International Conference on Language Resources and Evaluation (LREC'14), European Language Resources Association (ELRA), pp. 3145–3150. Reykjavik, Iceland (2014). http://www.lrec-conf.org/proceedings/lrec2014/pdf/1029_Paper.pdf

Khashabi, D., Sammons, M., Zhou, B., Redman, T., Christodoulopoulos, C., Srikumar, V., Rizzolo, N., Ratinov, L., Luo, G., Do, Q., Tsai, C.-T., Roy, S., Mayhew, S., Feng, Z., Wieting, J., Yu, X., Song, Y., Gupta, S., Upadhyay, S., Arivazhagan, N., Ning, Q., Ling, S., Roth, D.: CogCompNLP: your Swiss army knife for NLP. In: Proceedings of the Eleventh International Conference on Language Resources and Evaluation (LREC 2018). European Language Resources Association (ELRA). Miyazaki, Japan (2018). https://aclanthology.org/L18-1086

Kim, S., Toutanova, K., Yu, H.: Multilingual named entity recognition using parallel data and metadata from Wikipedia. In: Proceedings of the 50th Annual Meeting of the Association for Computational Linguistics (Volume 1: Long Papers), Association for Computational Linguistics, pp. 694–702. Jeju Island, Korea (2012). https://aclanthology.org/P12-1073

Kolitsas, N., Ganea, O.-E., Hofmann, T.: End-to-end neural entity linking. In: Proceedings of the 22nd Conference on Computational Natural Language Learning, Association for Computational Linguistics, pp. 519–529. Brussels, Belgium (2018). https://doi.org/10.18653/v1/K18-1050, https://aclanthology.org/K18-1050

Lample, G., Ballesteros, M., Subramanian, S., Kawakami, K., Dyer, C.: Neural architectures for named entity recognition. In: Proceedings of the 2016 Conference of the North American Chapter of the Association for Computational Linguistics: Human Language Technologies, Association for Computational Linguistics, pp. 260–270. San Diego, California (2016). https://doi.org/10.18653/v1/N16-1030, https://aclanthology.org/N16-1030

Lample, G., Conneau, A.: Cross-lingual language model pretraining. arXiv preprint arXiv:1901.07291 (2019)

Lewis, M., Liu, Y., Goyal, N., Ghazvininejad, M., Mohamed, A., Levy, O., Stoyanov, V., Zettlemoyer, L.: BART: denoising sequence-to-sequence pre-training for natural language generation, translation, and comprehension. In: Proceedings of the 58th Annual Meeting of the Association for Computational Linguistics, Association for Computational Linguistics, pp. 7871–7880, Online (2020). https://doi.org/10.18653/v1/2020.acl-main.703, https://aclanthology.org/2020.acl-main.703

Li, J., Sun, A., Han, J., Li, C.: A survey on deep learning for named entity recognition. IEEE Trans. Knowl. Data Eng. (2020b)

Lin, Y., Costello, C., Zhang, B., Lu, D., Ji, H., Mayfield, J., McNamee, P.: Platforms for non-speakers annotating names in any language. In: Proceedings of ACL 2018, System Demonstrations, Association for Computational Linguistics, pp. 1–6. Melbourne, Australia (2018). https://doi.org/10.18653/v1/P18-4001, https://aclanthology.org/P18-4001

Ling, X., Singh, S., Weld, D.S.: Design challenges for entity linking. Trans. Assoc. Comput. Linguistics **3**, 315–328 (2015)

Liu, Y., Ott, M., Goyal, N., Du, J., Joshi, M., Chen, D., Levy, O., Lewis, M., Zettlemoyer, L., Stoyanov, V.: RoBERTa: a robustly optimized BERT pretraining approach. arXiv preprint arXiv:1907.11692, abs/1907.11692 (2019)

Liu, Y., Gu, J., Goyal, N., Li, X., Edunov, S., Ghazvininejad, M., Lewis, M., Zettlemoyer, L.: Multilingual denoising pre-training for neural machine translation. Trans. Assoc. Comput. Linguistics **8**, 726–742 (2020)

Luo, G., Huang, X., Lin, C.-Y., Nie, Z.: Joint entity recognition and disambiguation. In: Proceedings of the 2015 Conference on Empirical Methods in Natural Language Processing, Association for Computational Linguistics, pp. 879–888. Lisbon, Portugal (2015). https://doi.org/10.18653/v1/D15-1104, https://aclanthology.org/D15-1104

Luong, T., Pham, H., Manning, C.D.: Bilingual word representations with monolingual quality in mind. In: Proceedings of the 1st Workshop on Vector Space Modeling for Natural Language Processing, Association for Computational Linguistics, pp. 151–159. Denver, Colorado (2015). https://doi.org/10.3115/v1/W15-1521, https://aclanthology.org/W15-1521

Ma, X., Hovy, E.: End-to-end sequence labeling via bi-directional LSTM-CNNs-CRF. In: Proceedings of the 54th Annual Meeting of the Association for Computational Linguistics (Volume 1: Long Papers), Association for Computational Linguistics, pp. 1064–1074, Berlin, Germany (2016). https://doi.org/10.18653/v1/P16-1101, https://aclanthology.org/P16-1101

Manning, C., Surdeanu, M., Bauer, J., Finkel, J., Bethard, S., McClosky, D.: The Stanford CoreNLP natural language processing toolkit. In: Proceedings of 52nd Annual Meeting of the Association for Computational Linguistics: System Demonstrations, Association for Computational Linguistics, pp. 55–60, Baltimore, Maryland (2014). https://doi.org/10.3115/v1/P14-5010, https://aclanthology.org/P14-5010

Mao, X., Dong, Y., He, S., Bao, S., Wang, H.: Chinese word segmentation and named entity recognition based on conditional random fields. In: Proceedings of the Sixth SIGHAN Workshop on Chinese Language Processing (2008). https://aclanthology.org/I08-4013

Mayhew, S., Roth, D.: TALEN: tool for annotation of low-resource ENtities. In: Proceedings of ACL 2018, System Demonstrations, Association for Computational Linguistics, pp. 80–86. Melbourne, Australia (2018). https://doi.org/10.18653/v1/P18-4014, https://aclanthology.org/P18-4014

Mayhew, S., Tsai, C.-T., Roth, D.: Cheap translation for cross-lingual named entity recognition. In: Proceedings of the 2017 Conference on Empirical Methods in Natural Language Processing, Association for Computational Linguistics, pp. 2536–2545. Copenhagen, Denmark (2017). https://doi.org/10.18653/v1/D17-1269, https://aclanthology.org/D17-1269

McCallum, A., Li, W.: Early results for named entity recognition with conditional random fields, feature induction and web-enhanced lexicons. In: Proceedings of the Seventh Conference on Natural Language Learning at HLT-NAACL 2003, pp. 188–191 (2003). https://aclanthology.org/W03-0430

McDonald, R., Petrov, S., Hall, K.: Multi-source transfer of delexicalized dependency parsers. In: Proceedings of the 2011 Conference on Empirical Methods in Natural Language Processing, Association for Computational Linguistics, pp. 62–72. Edinburgh, Scotland, UK (2011). https://aclanthology.org/D11-1006

McNamee, P., Mayfield, J.: Entity extraction without language-specific resources. In: COLING-02: The 6th Conference on Natural Language Learning 2002 (CoNLL-2002) (2002). https://aclanthology.org/W02-2020

Mihalcea, R., Csomai, A.: Wikify!: linking documents to encyclopedic knowledge. In: Proceedings of the ACM Conference on Information and Knowledge Management (CIKM) (2007)

Mikolov, T., Chen, K., Corrado, G., Dean, J.: Efficient estimation of word representations in vector space. In: Proceedings of the International Conference on Learning Representations (2013a)

Mikolov, T., Sutskever, I., Chen, K., Corrado, G.S., Dean, J.: Distributed representations of words and phrases and their compositionality. In: Proceedings of the Conference on Advances in Neural Information Processing Systems (NIPS) (2013c)

Milne, D., Witten, I.H.: Learning to link with Wikipedia. In: Proceedings of the ACM Conference on Information and Knowledge Management (CIKM) (2008)

Nadeau, D., Sekine, S.: A survey of named entity recognition and classification. Lingvisticae Investigationes **30**(1), 3–26 (2007)

Nguyen, D.B., Theobald, M., Weikum, G.: J-NERD: joint named entity recognition and disambiguation with rich linguistic features. Trans. Assoc. Comput. Linguistics **4**, 215–229 (2016)

Nie, Y., Tian, Y., Wan, X., Song, Y., Dai, B.: Named entity recognition for social media texts with semantic augmentation. In: Proceedings of the 2020 Conference on Empirical Methods in Natural Language Processing (EMNLP), Association for Computational Linguistics, pp. 1383–1391. Online (2020). https://doi.org/10.18653/v1/2020.emnlp-main.107, https://aclanthology.org/2020.emnlp-main.107

Nothman, J., Curran, J.R., Murphy, T.: Transforming Wikipedia into named entity training data. In: Proceedings of the Australasian Language Technology Association Workshop 2008, pp. 124–132. Hobart, Australia (2008). https://aclanthology.org/U08-1016

Nothman, J., Murphy, T., Curran, J.R.: Analysing Wikipedia and gold-standard corpora for NER training. In: Proceedings of the 12th Conference of the European Chapter of the ACL (EACL 2009), Association for Computational Linguistics, pp. 612–620. Athens, Greece (2009). https://aclanthology.org/E09-1070

Nothman, J., Ringland, N., Radford, W., Murphy, T., Curran, J.R.: Learning multilingual named entity recognition from Wikipedia. Artif. Intell. **194**, 151–175 (2013)

Pan, X., Zhang, B., May, J., Nothman, J., Knight, K., Ji, H.: Cross-lingual name tagging and linking for 282 languages. In: Proceedings of the 55th Annual Meeting of the Association for Computational Linguistics (Volume 1: Long Papers), Association for Computational Linguistics, pp. 1946–1958. Vancouver, Canada (2017). https://doi.org/10.18653/v1/P17-1178, https://aclanthology.org/P17-1178

Peng, N., Dredze, M.: Improving named entity recognition for Chinese social media with word segmentation representation learning. In: Proceedings of the 54th Annual Meeting of the Association for Computational Linguistics (Volume 2: Short Papers), Association for Computational Linguistics, pp. 149–155. Berlin, Germany (2016). https://doi.org/10.18653/v1/P16-2025, https://aclanthology.org/P16-2025

Peng, N., Dredze, M.: Named entity recognition for Chinese social media with jointly trained embeddings. In: Proceedings of the 2015 Conference on Empirical Methods in Natural Language Processing, Association for Computational Linguistics, pp. 548–554. Lisbon, Portugal (2015). https://doi.org/10.18653/v1/D15-1064, https://aclanthology.org/D15-1064

Pennington, J., Socher, R., Manning, C.: GloVe: global vectors for word representation. In: Proceedings of the 2014 Conference on Empirical Methods in Natural Language Processing (EMNLP), Association for Computational Linguistics, pp. 1532–1543. Doha, Qatar (2014). https://doi.org/10.3115/v1/D14-1162, https://aclanthology.org/D14-1162

Peters, M.E., Neumann, M., Iyyer, M., Gardner, M., Clark, C., Lee, K., Zettlemoyer, L.: Deep contextualized word representations. In: Proceedings of the 2018 Conference of the North American Chapter of the Association for Computational Linguistics: Human Language Technologies, Volume 1 (Long Papers), Association for Computational Linguistics, pp. 2227–2237. New Orleans, Louisiana (2018). https://doi.org/10.18653/v1/N18-1202, https://aclanthology.org/N18-1202

Raffel, C., Shazeer, N., Roberts, A., Lee, K., Narang, S., Matena, M., Zhou, Y., Li, W., Liu, P.J., et al.: Exploring the limits of transfer learning with a unified text-to-text transformer. The J. Machine Learn. Res. 21(140), 1–67 (2020)

Ratinov, L., Roth, D.: Design challenges and misconceptions in named entity recognition. In: Proceedings of the Thirteenth Conference on Computational Natural Language Learning (CoNLL-2009), Association for Computational Linguistics, pp. 147–155. Boulder, Colorado (2009). https://aclanthology.org/W09-1119

Reimersm, N., Gurevych, I.: Reporting score distributions makes a difference: performance study of LSTM-networks for sequence tagging. In: Proceedings of the 2017 Conference on Empirical Methods in Natural Language Processing, Association for Computational Linguistics, pp. 338–348. Copenhagen, Denmark (2017). https://doi.org/10.18653/v1/D17-1035, https://aclanthology.org/D17-1035

Rijhwani, S., Zhou, S., Neubig, G., Carbonell, J.: Soft gazetteers for low-resource named entity recognition. In: Proceedings of the 58th Annual Meeting of the Association for Computational Linguistics, Association for Computational Linguistics, pp. 8118–8123, Online (2020). https://doi.org/10.18653/v1/2020.acl-main.722, https://aclanthology.org/2020.acl-main.722

Sarawagi, S., Cohen, W.W.: Semi-Markov conditional random fields for information extraction. In: Proceedings of the Conference on Advances in Neural Information Processing Systems (NIPS) (2004)

Şeker, G.A., Eryiğit, G.: Initial explorations on using CRFs for Turkish named entity recognition. In: Proceedings of COLING 2012, The COLING 2012 Organizing Committee, pp. 2459–2474. Mumbai, India (2012). https://aclanthology.org/C12-1150

Sil, A., Yates, A.: Re-ranking for joint named-entity recognition and linking. In: Proceedings of the 22nd ACM International Conference on Information & Knowledge Management (CIKM), pp. 2369–2374. ACM (2013)

Strubell, E., Verga, P., Belanger, D., McCallum, A.: Fast and accurate entity recognition with iterated dilated convolutions. In: Proceedings of the 2017 Conference on Empirical Methods in Natural Language Processing, Association for Computational Linguistics, pp. 2670–2680. Copenhagen, Denmark (2017). https://doi.org/10.18653/v1/D17-1283, https://aclanthology.org/D17-1283

Täckström, O., McDonald, R., Uszkoreit, J.: Cross-lingual word clusters for direct transfer of linguistic structure. In: Proceedings of the 2012 Conference of the North American Chapter of the

Association for Computational Linguistics: Human Language Technologies, Association for Computational Linguistics, pp. 477–487. Montréal, Canada (2012). https://aclanthology.org/N12-1052

Tjong Kim Sang, E.F., De Meulder, F.: Introduction to the CoNLL-2003 shared task: Language-independent named entity recognition. In: Proceedings of the Seventh Conference on Natural Language Learning at HLT-NAACL 2003, pp. 142–147 (2003). https://aclanthology.org/W03-0419

Tjong Kim Sang, E.F.: Introduction to the CoNLL-2002 shared task: language-independent named entity recognition. In: COLING-02: The 6th Conference on Natural Language Learning 2002 (CoNLL-2002) (2002). https://aclanthology.org/W02-2024

Tsai, C.-T., Mayhew, S., Roth, D.: Cross-lingual named entity recognition via wikification. In: Proceedings of the 20th SIGNLL Conference on Computational Natural Language Learning, Association for Computational Linguistics, pp. 219–228. Berlin, Germany (2016). https://doi.org/10.18653/v1/K16-1022, https://aclanthology.org/K16-1022

Tsai, C.-T., Roth, D.: Cross-lingual wikification using multilingual embeddings. In: Proceedings of the 2016 Conference of the North American Chapter of the Association for Computational Linguistics: Human Language Technologies, Association for Computational Linguistics, pp. 589–598. San Diego, California (2016b). https://doi.org/10.18653/v1/N16-1072, https://aclanthology.org/N16-1072

Tür, G.: A statistical information extraction system for Turkish. Ph.D. thesis, Bilkent University (2000)

Wang, M., Manning, C.D.: Cross-lingual projected expectation regularization for weakly supervised learning. Transa. Assoc. Comput. Linguistics **2**, 55–66 (2014)

Weischedel, R., Palmer, M., Marcus, M., Hovy, E., Pradhan, S., Ramshaw, L., Xue, N., Taylor, A., Kaufman, J., Franchini, M., et al.: Ontonotes release 5.0. In: Linguistic Data Consortium, Philadelphia, PA, vol. 23 (2013)

Xue, L., Constant, N., Roberts, A., Kale, M., Al-Rfou, R., Siddhant, A., Barua, A., Raffel, C.: mT5: a massively multilingual pre-trained text-to-text transformer. In: Proceedings of the 2021 Conference of the North American Chapter of the Association for Computational Linguistics: Human Language Technologies, Association for Computational Linguistics, pp. 483–498. Online (2021). https://doi.org/10.18653/v1/2021.naacl-main.41, https://aclanthology.org/2021.naacl-main.41

Yadav, V., Bethard, S.: A survey on recent advances in named entity recognition from deep learning models. In: Proceedings of the 27th International Conference on Computational Linguistics, Association for Computational Linguistics, pp. 2145–2158. Santa Fe, New Mexico, USA (2018). https://aclanthology.org/C18-1182

Yarowsky, D., Ngai, G., Wicentowski, R.: Inducing multilingual text analysis tools via robust projection across aligned corpora. In: Proceedings of the First International Conference on Human Language Technology Research (2001). https://aclanthology.org/H01-1035

Yarowsky, D., Ngai, G.: Inducing multilingual POS taggers and NP bracketers via robust projection across aligned corpora. In: Second Meeting of the North American Chapter of the Association for Computational Linguistics (2001). https://aclanthology.org/N01-1026

Z. Ye and Ling, Z.-H.: Hybrid semi-Markov CRF for neural sequence labeling. In: Proceedings of the 56th Annual Meeting of the Association for Computational Linguistics (Volume 2: Short Papers), Association for Computational Linguistics, pp. 235–240. Melbourne, Australia (2018). https://doi.org/10.18653/v1/P18-2038, https://aclanthology.org/P18-2038

Yeniterzi, R.: Exploiting morphology in Turkish named entity recognition system. In: Proceedings of the ACL 2011 Student Session, Association for Computational Linguistics, pp. 105–110. Portland, OR, USA (2011). https://aclanthology.org/P11-3019

Zeman, D., Resnik, P.: Cross-language parser adaptation between related languages. In: Proceedings of the IJCNLP-08 Workshop on NLP for Less Privileged Languages (2008). https://aclanthology.org/I08-3008

Zhang, S., Qin, Y., Wen, J., Wang, X.: Word segmentation and named entity recognition for SIGHAN bakeoff3. In: Proceedings of the Fifth SIGHAN Workshop on Chinese Language Processing, Association for Computational Linguistics, pp. 158–161. Sydney, Australia (2006). https://aclanthology.org/W06-0126

Zhang, Y., Yang, J.: Chinese NER using lattice LSTM. In: Proceedings of the 56th Annual Meeting of the Association for Computational Linguistics (Volume 1: Long Papers), Association for Computational Linguistics, pp. 1554–1564. Melbourne, Australia (2018). https://doi.org/10.18653/v1/P18-1144, https://aclanthology.org/P18-1144

Zhu, Y., Wang, G.: CAN-NER: convolutional Attention Network for Chinese Named Entity Recognition. In: Proceedings of the 2019 Conference of the North American Chapter of the Association for Computational Linguistics: Human Language Technologies, Volume 1 (Long and Short Papers), Association for Computational Linguistics, pp. 3384–3393. Minneapolis, Minnesota (2019). https://doi.org/10.18653/v1/N19-1342, https://aclanthology.org/N19-1342

Zhuo, J., Cao, Y., Zhu, J., Zhang, B., Nie, Z.: Segment-level sequence modeling using gated recursive semi-Markov conditional random fields. In: Proceedings of the 54th Annual Meeting of the Association for Computational Linguistics (Volume 1: Long Papers), Association for Computational Linguistics, pp. 1413–1423. Berlin, Germany (2016). https://doi.org/10.18653/v1/P16-1134, https://aclanthology.org/P16-1134

Identifying Entity Candidates 5

In Chap. 3, we saw that a typical knowledge base can contain millions of entities. At inference time, considering all entities in a knowledge base as possible disambiguation for a mention is usually impractical especially for more expressive models. Therefore, it is prudent to quickly filter out irrelevant entities that are unlikely disambiguation for a mention. For instance, it is unlikely that a mention "Chicago" would refer to the entity `George_Bush`. The candidate generation step helps organize the search space of entities and provides useful features that can be used at inference time. A good candidate generation method may not only speed up the pipeline but also improve disambiguation accuracy. Figure 5.1 illustrates the input and output of this step. In this example, two candiadte entities, `Alex_Smith` and `Alex_Smith_(athlete)`, are retrieved for the mention "Alex Smith". Given the extracted mentions in text, a candidate generation algorithm retrieves a short list of entries from the knowledge base for each mention.

More formally, a candidate generation algorithm identifies a small proper subset $\mathcal{C}(m) \subsetneq \mathcal{K}$ of plausible entities in the knowledge base $\mathcal{K}$ that are possible disambiguation for a mention m. Given m, candidate generation outputs a list of candidate entities $\mathcal{C}(m) = \{e_1, e_2, \ldots, e_k\}$ of size at most k, each associated with a prior probability $\Pr_{\text{prior}}(e_i \mid m)$ indicating the probability of m referring to e_i, given only m's surface string. The candidates are ranked according to this prior probability. Only the top k entities will be passed on to the following inference step.

Candidate generation is critical to the overall performance of entity linking (Hachey et al. 2013). There is a natural trade-off of the size of candidate sets and inference speed. On one hand, if we only generate few candidates, the correct disambiguation might be lost. On the other hand, too many candidates will make the inference problem in the next step harder and slower. Therefore, the goal of this step is to quickly generate a small set of candidate entities that the mention may refer to. To achieve quick retrieval, most candidate generation

Fig. 5.1 An example of input and output of the candidate generation step. For each extracted mentions in the input text, an algorithm in this step will retrieve a list of entity candidates from the knowledge base

approaches resort to dictionary based methods or simple name similarity based methods. That is, no contextual information in the query document is used in this step. However, with the growth of computational power and better similarity searching algorithms, some context-sensitive candidate generation approaches have been proposed as well.

In this chapter, we will discuss various candidate generation approaches that have been used in monolingual and cross-lingual entity linking, their limitations, and future directions worthy exploring to improve candidate generation.

5.1 Monolingual Candidate Generation

We start by introducing monolingual candidate generation approaches, in which the input text and the knowledge base are in the same language. Both monolingual and cross-lingual methods can be categorized into three broad categories: dictionary based, query-log based, and retrieval based approaches.

5.1.1 Dictionary Based Approaches

As discussed earlier, traditional candidate generation approaches ignore the context of the mention in the query document. This naturally leads to a dictionary lookup based approach to candidate generation.

A dictionary lookup based method collects as many as possible names that each entity may be referred to, and use this information to construct a dictionary. More specifically, each entry in the dictionary is a (*key*, *value*) pair, where the *key* is a string and the corresponding *value* contains all the possible entities which can be referred by the string. For instance, the string "Chicago" may have the following entities as the values: `Chicago`, `Chicago_(magazine)`, `Chicago_(band)`, `Chicago_Park,_California`, and so on.

5.1 Monolingual Candidate Generation

Table 5.1 Part of a dictionary compiled from Wikipedia hyperlinks showing the entities that the string *chicago* can refer to, along with the respective prior probabilities and counts

String	Entity (e)	$\Pr_{\text{prior}}(e \mid m)$	Counts/total
chicago	Chicago	0.275	62,939/228,690
chicago	University_of_Chicago	0.075	17,186/228,690
chicago	Chicago_Cubs	0.045	10,332/228,690
chicago	Chicago_Tribune	0.041	9,457/228,690
...	Chicago_White_Sox	0.040	9,184/228,690
...	Chicago_Bears	0.039	9,018/228,690
...	Art_Institute_of_Chicago	0.015	3,538/228,690
...	Chicago_(band)	0.005	1,225/228,690
chicago	Loyola_University_Chicago	0.005	1,057/228,690

5.1.1.1 Constructing Dictionary

To do this, first a dictionary is compiled that maps mention surfaces strings to all possible entities that they can refer to. If the target KB is Wikipedia, one common approach to compile such a dictionary is to crawl a large collection of hyperlinked documents and computing the frequency with which an anchor text links to a page in Wikipedia, such as the one shown in Table 5.1. From the frequencies, one can estimate the conditional probability $\Pr_{\text{prior}}$. For instance, if the string *chicago* appears hyperlinked 228,690 times in the document collection, and links to Chicago and University_of_Chicago 62,939 and 17,186 times respectively, then the probability of a mention "Chicago" referring to Chicago and University_of_Chicago is 0.275 and 0.075 respectively.

In the literature, the collection of documents used to compile this dictionary can either be hyperlinked text from the Web (Spitkovsky and Chang 2012), or simply the articles in Wikipedia itself (Ratinov et al. 2011). Wikipedia has emerged as a popular source for such candidate generation, owing to the rich metadata available for each entity. The following information in Wikipedia has been used to construct the dictionary for generating entity candidates.

- **Title**: The title of a Wikipedia article can be included as a key for the entity.
- **Anchor Text**: The hyperlinked phrases in Wikipedia articles could provide name variations for Wikipedia entries. More specifically, the hyperlinked phrase is the key, and the linked Wikipedia entity is the value.

- **Redirects**: For example, if the string "US" is redirected to Wikipedia page `United_States`, "US" will be considered as one of the keys for the entiyt `United_States`.
- **Disambiguation Pages**: The entities contained in a Wikipedia disambiguation page can be used as the values of the target string which is being disambiguated.
- **Bold Phrases**: In the first paragraph of Wikipedia articles, sometimes there are bold-faced phrases which are the aliases of the entity. For instance, in the page of `United_States`, "United State of America", "USA", "United States", "U.S.", and "America" are all bold-faced phrases in the very first sentence. These phrases can be used as the keys for the entity `United_States`.

In addition to Wikipedia, CrossWiki (Spitkovsky and Chang 2012) has become a popular resource for candidate generation in entity linking (Ling et al. 2015; Ganea et al. 2016). CrossWiki is a pre-compiled candidate dictionary, built by crawling the web to find hyper-linked phrases which point to some Wikipedia pages. This dictionary contains more than 175 million unique keys (strings) along with the Wikipedia entities they may represent.

5.1.1.2 Using The Dictionary

Given a mention, the simplest way to generate candidate entities is by exactly matching the mention surface string with the keys in the dictionary. If there is a match, the entities in the corresponding value set are taken as the candidates. Besides performing exact match, some systems (Dredze et al. 2010; Tsai and Roth 2016a; Tsai and Roth 2016b) use partial match. For instance, the keys that are wholly contained in the mention, or the keys have strong string similarity scores with the mention are all considered matched. The similarity measures could be character Dice score, skip bigram Dice score, or Hamming distance. For all the matched keys, the merged value sets are used as the candidates.

Before looking up the dictionary, some approaches try to correct misspellings in the mentions. This is particularly useful for text which has not been carefully edited such as documents from discussion forums, weblogs, and social medias. Zhang et al. (2010) use the "Did you mean" feature in Wikipedia search to correct misspellings. For example, if we search "Abbot Nutrition" in Wikipedia, the first sentence of the result page is "Did you mean: abbott nutrition", which corrects the misspelled "Abbot". Similarly, Zheng et al. (2010) use query spelling correction of Google search engine to correct the misspellings in the mentions.

Simple coreference resolution or surface form expansion is sometimes applied to resolve the short mentions before querying the dictionary (Cucerzan 2007). Since short mentions are usually more ambiguous (e.g., acronyms or last names), the idea is that the corresponding full names could be mentioned somewhere in the same document. Most approaches use heuristic rules to match adjacent mentions (e.g., "University of Illinois at Urbana-Champaign (UIUC)"), or to match mentions in the entire document (e.g., "George W. Bush" may be mentioned before "Bush" in the same document). Zhang et al. (2011) propose to learn a

5.1 Monolingual Candidate Generation

supervised classifier to decide if a mention could be an acronym of another mention for more challenging cases. For example, "CCP" stands for "Communist Party of China", and "MOD"/"MINDEF"/"MD" either of which can stand for "Ministry of Defense". They not only look at the mention string, but also consider neighboring words of the mentions. They show substantial improvement over rule-based methods.

The retrieved candidate set could be very large for some highly ambiguous mentions, say, contains more than 100 entities. Especially when the partial match methods are used. This large set of candidates may make the later inference problem difficult and inefficient. Therefore, most entity linking systems only keep top k candidates from the full candidates set, where k is usually less than 30. This ranking of candidates is usually based on some popularity-based measures. A common approach is to compute prior probabilities, Pr(entity|mention), from Wikipedia anchor text:

$$\Pr(\text{entity}|\text{mention}) = \frac{\text{\# times the mention is linked to the entity}}{\text{\# times the mention is hyperlinked}} \quad (5.1)$$

From the examples in Table 5.1, we can see that the prior probabilities Pr(Chicago|*chicago*) = 0.275 and Pr(Chicago_Cubs|*chicago*) = 0.075.

5.1.2 Query Log Based Approaches

Similar to Wikipedia, an encyclopedia collectively created by its users, query log is another resource which is generated by users of search engines collaboratively. As candidate generation can be viewed as a retrieval problem, search engines have been used for retrieving possible candidate entities.

For instance, Dredze et al. (2010) leverage a search engine to query candidates for entity linking. They query the mention string on Google Search and collect Wikipedia pages within the top 20 returned documents as the candidates. Moreover, the rank of a candidate's Wikipedia page in a Google query is included as a popularity feature in their model.

Besides using Google Search, Zhang et al. (2010) query Wikipedia search for infrequent mentions. If other candidate generation methods fail to retrieve any entity, they query the mention in Wikipedia search. The top-1 returned Wikipedia page will be included as a candidate if the similarity between the page title and the mention is higher than a pre-specified threshold value.

5.1.3 Retrieval Based Approaches

Dictionary-based method is very popular because of its simplicity and efficiency. However, the performance of dictionary-based method depends on the coverage of the observed mention-entity pairs which are used in constructing the dictionary. When the target knowledge base is Wikipedia, this method might be sufficient since there are lots of mention-entity pairs in Wikipedia articles or on the Web. Namely, many articles contain entities or concepts that are linked to Wikipedia entries. For emerging entities, specific domains, or other knowledge bases, dictionary based method may not perform as well. In such cases, a retrieval based approach is needed.

A retrieval based approach compares the mention with all entities in the knowledge base, and retrieves top k similar entities. It is usually less efficient than the dictionary based approach when the target knowledge base is sufficiently large. However, with proper indexing and searching techniques, fast retrieval can still be achieved. Based on whether the document context is used in computing the similarity metric, we have two categories of approaches: context-insensitive similarity and context-sensitive similarity.

5.1.3.1 Context-Insensitive Similarity

Logeswaran et al. (2019) propose a zero-shot entity linking task. They construct a dataset consists of 16 domains from FANDOM.[1] Each domain is a fictional universe from a book or film series. The task is zero-shot in the sense that no test entities have been mentioned in the training documents. They further assume that each entity in the knowledge base only have a text description. Therefore, there is no alias table or other structured data for constructing the dictionary for candidate generation.

They propose to use BM25 (Robertson and Walker 1994) to retrieve candidate entities. In information retrieval, BM25 is a ranking function used by search engines to estimate the relevance between a given search query and documents. It is a TF-IDF-like retrieval function that ranks a set of documents based on query terms appearing in the document. In the entity linking setup, a mention string is analogous to the query and entity descriptions are the documents. They leverage the implementation in Lucene.[2] In their experiments, the coverage of top-64 candidates is less than 77% on average, which indicates the difficulty of the proposed dataset.

5.1.3.2 Context-Sensitive Similarity

Neural network models have been proposed to encode mentions and entities for candidate generation. Gillick et al. (2019) propose a dual-encoder model to encode mentions and entities in the same dense vector space. Figure 5.2 from the original paper shows the model architecture. The mention encoder takes a mention span, its left context, its right context, and the sentence as input. The entity encoder uses entity title, the first paragraph of Wikipedia article, and Wikipedia categories. Both encoders use feed-forward neural networks with pretrained GloVe word embeddings. Finally, cosine similarity is used as the similarity measure

---

[1] https://www.fandom.com/.
[2] https://lucene.apache.org/.

5.1 Monolingual Candidate Generation

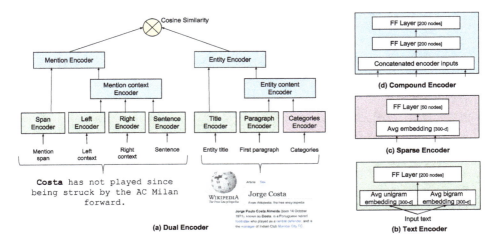

Fig. 5.2 The model architecture of the candidate retrieval method proposed in Gillick et al. (2019)

between the encoded representations of mention and entity. The model is trained on anchor texts in Wikipedia with an approach that adds hard negative examples.

Once the model is trained, the entity encoder is used to pre-compute representations of all entities in the knowledge base. At inference time, mentions are encoded by the mention encoder and candidate entities are retrieved based on their cosine similarity.

We note that different from the previously discussed candidate generation approaches which are mostly context-insensitive, this model does incorporate mention context. The authors find that using the top-1 retrieved entity as the answer without further re-ranking has already outperformed several other entity linking models. The idea of encoding mentions and entities separately have been explored in earlier works such as Gupta et al. (2017). However, Gupta et al. (2017) still use a dictionary based approach to generate candidates quickly. The mention and entities representations are only used in the following re-ranking step.

Similar to the dual encoder idea, Wu et al. (2020) propose a bi-encoder model for candidate generation. Instead of using feed-forward neural networks, they use two independent BERT transformer models to encode mention and entity. The input to the mention encoder is "[CLS] ctxt$_l$ [M$_s$] mention [M$_e$] ctxt$_r$ [SEP]", where ctxt$_l$ and ctxt$_r$ are the word-pieces tokens of the left and right context, and [M$_s$] and [M$_e$] are special tokens to tag the start and end of the mention. The input to the entity encoder is "[CLS] title [ENT] description [SEP]", where title and description are word-pieces tokens of the entity title and this entity's description. For both mention and entity encoders, they use the "[CLS]" token from the last layer of transformer models as the representations. After using the bi-encoder model to retrieve top 100 candidate entities, they propose a more expressive cross-encoder model to re-rank the candidates. Figure 5.3 illustrates the idea behind the proposed bi-encoder and cross-encoder.

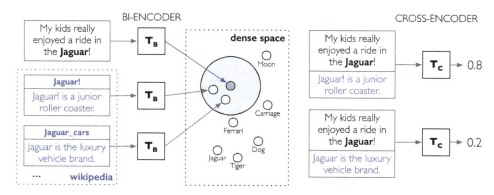

Fig. 5.3 An overview of the models proposed by Wu et al. (2020). The bi-encoder model is used as a candidate generator, whereas the cross-encoder is for re-ranking the retrieved candidate entities

In order to improve the efficiency of retrieving top k similar dense vectors, they utilize the exact and approximate nearest neighbor search algorithms implemented in FAISS (Johnson et al. 2019). In their experiments, an exact search method takes 9.2 ms on average to return top 100 candidates per query. An approximate search reduces the average query time to 2.6 ms with less than 1.2% drop in accuracy and recall.

5.2 Cross-Lingual Candidate Generation

The most common cross-lingual candidate generation setup is to retrieve English entity candidates given non-English mentions, since several English knowledge bases are much larger than the non-Enlgish ones. This is arguably the most important step in the cross-lingual entity linking pipeline as it is the first component that inferences between two languages. Unlike monolingual entity linking, in which dictionary-based methods can perform reasonably well, candidate generation in cross-lingual entity linking is much more challenging.

A naive approach for cross-lingual candidate generation is to query the English candidate dictionary directly using a foreign mention. But this approach only works if the target language is very close to English, so that names in the two languages are expressed almost exactly and in the same writing script. For instance, mentions of Barack Obama in Spanish text are usually exactly the same as English spelling. Querying the English dictionary directly would work well for such mentions. However, this approach does not work for most languages.

In the following sections, we will discuss candidate generation approaches that have been proposed in the cross-lingual entity linking literature. Similar to the monolingual candidate generation approaches, dictionary based, query log based, and retrieval based approaches have all been adapted to the cross-lingual scenario. In addition, we will discuss name transla-

5.2 Cross-Lingual Candidate Generation

tion based methods in the end which essentially attempt to convert the cross-lingual problem into monolingual problem.

5.2.1 Dictionary Based Approaches

Dictionary based approaches have enjoyed success in monolingual entity linking. Naturally, many systems have attempted to re-purpose them for cross-lingual entity linking (Tsai and Roth 2016b; Sil and Florian 2016).

In this approach, a candidate dictionary for monolingual entity linking in the target language is computed. For the example scenario in Fig. 5.4, this involves computing a dictionary from Russian strings to entities in the Russian Wikipedia. When generating candidates for a Russian mention, first dictionary lookup is used to generate candidates in the target language, as done in monolingual entity linking. That is, for the mention Берлин in Fig. 5.4, monolingual candidate generation identifies candidates in the Russian Wikipedia. Then, each entity in the produced candidate set is mapped to its corresponding entity in the English Wikipedia via inter-language links. These mapped English entities (`Berlin`, `Berlin,_New_Hamphshire`, and `Boris_Berlin`) are returned as the candidates.

This approach relies on the existence of a hyper-linked document collection in the target language and a mapping of entities in the target language to entities in English. Both these resources might not be available for low resource languages. More specifically, the quality of the Russian dictionary depends on both the quantity of hyperlinked text in the Russian Wikipedia articles and the size of the Russian Wikipedia. As we can see in Fig. 5.4, the Russian Wikipedia does not contain `Berlin_(TV_series)`. If the mention actually refers to this TV series, this approach would fail to retrieve the correct entity in the English Wikipedia. The second step that finds the corresponding English entries depends on the inter-language links between the two languages. In Fig. 5.4, we see that the Russian page Берлин, Анна Катарина is not linked to an English page. In fact, this entity does not exist in the English Wikipedia. Furthermore, even the entity exist in both Russian and

Fig. 5.4 Extending dictionary based approach for cross-lingual candidate generation

English Wikipedia, the inter-language link could still be missing. As a result, the correct disambiguation might not be present in the list of candidates produced using this approach.

5.2.2 Query Log Based Approaches

Similar to Wikipedia, an encyclopedia created by its users, query log can be viewed as a resource which is collaboratively generated by a large number of users of search engines. The size of query log is often orders of magnitude larger than Wikipedia. In fact, Wikipedia is a subset of query log since a search of non-English mention may lead to the corresponding English Wikipedia entity if there is an inter-language link between this entity's entries in different languages.

To overcome the limitations of dictionary based approach, Fu et al. (2020) propose a query log based approach for cross-lingual candidate generation. They search the morphologically normalized foreign mention in Google Search, and pick top k Wikipedia web-page results. If the results contain any target-language Wikipedia page which is linked to the corresponding English Wikipedia page via a inter-language link, they will add this English entity into the candidate set. Moreover, when the mention is a geopolitical or location entity, they also search in Google Map. The English location names returned by Google Map will then be searched in Google Search in order to get the possible Wikipedia entries.

Using this approach together with the dictionary based approach yields substantial improvement on candidate coverage for both low-resource and high-resource languages.

5.2.3 Retrieval Based Approaches

Instead of relying on other resources such as hyperlinked corpus, Wikipedia inter-language links, or query logs, retrieval based approaches directly compute similarity between mentions and entities of two different languages. Although the quality of this cross-lingual similarity score still largely depends on the availability of cross-lingual signals, these methods might work better for emerging entities or less structured knowledge bases. One drawback of retrieval based approaches is that they are usually less efficient than dictionary based approaches since a mention will be compared with all entities in a knowledge base.

In this section, we will introduce two retrieval-based candidate generation methods for cross-lingual EL. Instead of further re-ranking the top-retrieved candidates, both models return the entity with the highest similarity score as the predicted grounding for a mention.

5.2 Cross-Lingual Candidate Generation

5.2.3.1 Context-Insensitive Similarity

One approach for retrieving candidates is simply based on name similarity across two languages. It is context-insensitive in the sense that other words in the input document and other information about the entity in the KB are not used in computing this similarity metric.

Rijhwani et al. (2019) proposed a zero-shot cross-lingual candidate generation approach. Figure 5.5a illustrates the key components. For a high-resource language (Hindi in the example) which have enough entity pairs with English, an entity similarity model is proposed to encode entity names in the two languages. Two Bi-LSTM models are applied on the character embeddings to encode entity names of the two languages. These LSTMs are trained to score the entity pairs which are in the entity map higher than other random pairs.

To generate candidates for a low-resource language, they proposed a *Pivoting* idea. First, a related high-resource language is chosen. The chosen high-resource language (Hindi) should have the same orthography or phonology to the low-resource language (Marathi). Then the pre-trained entity encoder for this high-resource language is used to encode the low-resource mention directly. Figure 5.5b illustrates the process. The encoded low-resource mention, v_m, will be compared with both English entities and their counterparts in the chosen high-resource language.

A limitation of pivoting is that many languages often do not have a high resource relative. For instance, several languages have writing scripts that are either unique to that language (e.g., Sinhalese, Thai, Armenian) or belong to a language family in which most languages are low-resource (e.g., Tigrinya). In such cases, approaches like pivoting might not work well.

(a) Training an entity similarity model on "parallel" entities in Hindi and English. Vector representations v_{HRL} and v_{en} are trained so that parallel entities have high cosine similarity scores.

(b) Computing similarity score between English and Marathi entities, using pivoting through a high resource related language e_{HRL} (in this case Hindi with the same writing script).

Fig. 5.5 Pivoting for Marathi candidate generation. An entity similarity model between English and a high resource relative of Marathi (in this case Hindi with the same writing script) is trained (**a**). This similarity model is used to compute similarity between Marathi entity mentions and English entities in the KB for retrieval (**b**)

5.2.3.2 Context-Sensitive Similarity

The dual-encoder model that we saw in the monolingual candidate generation section has been adapted to the cross-lingual setup. Botha et al. (2020) proposed to encode mentions and entities by BERT-based Transformer network, which is initialized from a pre-trained multilingual BERT model. The inputs to the mention encoder include mention surface string, left context, right context, and document title, whereas the entity description is encoded by the entity encoder to generate the entity representation. The cosine similarity is used as the metric for scoring a mention-entity pair. This model is trained on anchor texts from 104 languages in Wikipedia.

5.2.4 Name Translation and Transliteration Based Approaches

Another type of approaches that is specific for cross-lingual candidate generation is to convert the problem into the monolingual candidate generation problem. Namely, if we can translate non-English mentions into English, any monolingual candidate generation methods can be applied on the translated mention.

This approach is usually used together with other cross-lingual candidate generation methods, thus results in a hybrid candidate generation approach. Figure 5.6 shows an example of a hybrid approach based on dictionary lookup methods. First, cross-lingual candidate generation is attempted using the approach described in the previous sections. If no candidates are found, the mention surface is translated or transliterated into English. Then English entity candidates are retrieved using the translated mention in some monolingual candidate generation method.

We can see that the key component of this approach is the name translation or transliteration module. In this section, we will first introduce the name transliteration problem with some popular models. We will then discuss how these techniques have been used for solving the cross-lingual entity linking problem.

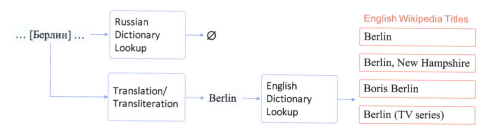

Fig. 5.6 A hybrid approach that combines a cross-lingual dictionary candidate generation method and name translation. By translating the non-English mention into English, a monolingual candidate generation method can be applied on the translated mention directly

5.2.4.1 Transliteration

Transliteration is the process of transducing a name from one writing system to another (e.g., ओबामा in Devanagari to *Obama* in Latin script) while preserving its pronunciation (Knight and Graehl 1997; Karimi et al. 2011).

Historically, the transliteration task emerged as a component of machine translation systems, to handle the translation of names and technical terms across languages. Such lexical items will have poor coverage in a bilingual dictionary or phrase translation table, thus making it necessary to replace them with their (approximate) phonetic equivalent. However, transliteration has utility beyond as a sub-routine in machine translation. For instance, transliterating names from foreign languages to English[3] helps in multilingual knowledge acquisition tasks like named entity recognition (Darwish 2013) and information retrieval (Virga and Khudanpur 2003; Jaleel and Larkey 2003).

Two tasks feature prominently in the transliteration literature: *generation* (Knight and Graehl 1997) that involves producing an appropriate transliteration for a given word in an open-ended way, and *discovery* (Sproat et al. 2006; Klementiev and Roth 2008) that involves selecting an appropriate transliteration for a word from a list of candidates.

5.2.4.2 Transliteration Generation

Generation aims to transliterate an input word $x = (x_1, x_2, \ldots, x_n)$ in the source writing script without relying on a translation lexicon, by generating a sequence of characters $y = (y_1, y_2, \ldots, y_m)$ in the target writing script. Figure 5.7a shows how a generation model will process the Hindi string ब्रसेल्स . A model directly outputs a sequence of characters in the target language.

Transliteration generation is often formulated as a *machine translation (MT) problem*, by treating the characters x_i in the input word as "words", the input word x as the "sentence", and the target output y as the "translation". This allows one to use available MT toolkits such as Moses (Koehn et al. 2007) or Joshua (Li et al. 2009) for training a transliteration system. However, these approaches may not exploit structure that is specific to the transliteration task.

(a) Generation   (b) Discovery

Fig. 5.7 Generation treats transliteration as a sequence transduction task, while discovery aims to select the correct transliteration from a given list of names

---

[3] Also referred to as *back-transliteration*.

Other approaches treat generation as a *sequence labeling problem*. Each character x_i in the input is assigned a label $t_i = y_p, \ldots, y_q$ which is a sequence of one or more characters from the target writing script. For instance, the Hindi character थ can be labeled as "tha" when transliterating Hindi→English. One issue with this formulation is that the size of output label space can be quite large, depending on how many different spans $y_p, \ldots, y_q$ are seen in the training data. For instance, Ammar et al. (2012) reported that Arabic → English transliteration had > 1200 labels, while Thai → English had > 1700 labels.

Both the above formulations require generous amount of name pairs ($\approx$ 5–10k) as supervision. One popular way to obtain this supervision is identifying name pairs using inter-language links in Wikipedia (Irvine et al. 2010). However, a truly low-resource language (like Tigrinya) is likely to have limited Wikipedia presence as well, and thus low resource transliteration remains a challenging problem.

5.2.4.3 Transliteration Discovery

On the other hand, the task of transliteration discovery deals with selecting the correct transliteration y for a word x from a relatively small list of candidates $\mathcal{N}$. The presence of the name list makes discovery essentially a *ranking* problem—rank all names in $\mathcal{N}$ to identify the most appropriate transliteration for a given word in the source script. Figure 5.7b shows how a discovery model will process the Hindi string ब्रसेल्स .

Discovery is a considerably easier task than generation, owing to the restricted search space. Indeed, for $|\mathcal{N}| \sim 50k$, unsupervised discovery approaches that use constraints derived from romanization tables have been shown to be successful (Chang et al. 2009). Notice that for a given input x, many names in $\mathcal{N}$ can be filtered out quite easily, so the effective search space is even smaller. In general, the hardness of discovery problem increases with $|\mathcal{N}|$, say, when $\mathcal{N}$ contains millions of names.

A key limitation of discovery is the assumption that the correct transliteration(s) is in the list of candidates $\mathcal{N}$. But it is unlikely that $\mathcal{N}$ will be exhaustive, as new names are constantly being introduced in language.[4] If no correct transliteration is present in $\mathcal{N}$, a discovery model will end up selecting *some* name from $\mathcal{N}$, and produce false positives. Of course, one can partially remedy this issue by using a large list of candidates $\mathcal{N}$, at the cost of increased difficulty.

Another issue with transliteration discovery approaches is that they often exploit features derived from resources that are unlikely to be available for low-resource languages, like temporally aligned comparable corpora (Sproat et al. 2006; Klementiev and Roth 2008). To overcome these limitations, it is prudent to develop generation models that can handle input for which the transliteration does not belong in $\mathcal{N}$, and operate in low-resource scenarios.

---

[4] The honorific *Khaleesi* from the TV series "Game of Thrones" became a popular baby name in 2016.

5.2.4.4 Name Translation for Cross-Lingual Entity Linking

In this section, we introduce two works that use name translation for cross-lingual entity linking problem. Tsai and Roth (2018) propose a probabilistic name translation model for cross-lingual entity linking. The traditional setting of transliteration focuses only on single-token names of people or locations, but for entity linking, the entity names are often longer (e.g., names of organizations). Moreover, multi-token names of locations and organizations typically require a mixture of translation and transliteration.

The proposed model is learnt from Wikipedia title pairs obtained from the inter-language links in Wikipedia. Since the model is learnt from phrase pairs rather than word pairs, they extend a transliteration model to jointly model word alignment and word-to-word transliteration. The key idea is that if words in a phrase pair can be aligned well, a word transliteration model can learn from cleaner signals. On the other hand, a good transliteration model can help to improve word alignment performance, because the transliteration model may provide better word generation probability if the word pair appears infrequently in the training data, but the sub-words pairs are frequent enough. That is, word transliteration and alignment models could reinforce each other, therefore yielding a better result.

More specifically, given a Wikipedia title pair (F, E), where F is the foreign title and E is the target English title, let the number of words in F and E be m and l respectively, the title generation probability is

$$P(F, A|E, m) = P(A|m) \prod_{(f,e) \in A} P(f|e), \qquad (5.2)$$

where A is an alignment assignment of words $f \in F$ and $e \in E$. The alignment A is a list of size $|E| = l$, where $A[j]$ could either be null, or the index of the word in F which is aligned with the j-th target word. Figure 5.8 shows two examples. Given a Spanish-English title pair ("universidad de keiō", "keio university"), the word alignments variable A in the left example is [2, 0]. The 2 in the first position means that "keio" is aligned with "keiō", and the 0 indicates "university" is aligned with "universidad". In the right example, since the word "university" is not aligned with any source word, the second element in A becomes *null*.

The last term of Eq. (5.2) is the word generation probability given the word alignment, where $(f, e) \in A$ is a word pair according to the alignment A. In the left example of Fig. 5.8, there are two word pairs, (universidad, university) and (keiō, keio), therefore $\prod_{(f,e) \in A} P(f|e) = P(\text{universidad}|\text{university})P(\text{keiō}|\text{keio})$. This is where a transliteration model (Pasternack and Roth 2009) is used to estimate word generation probability.

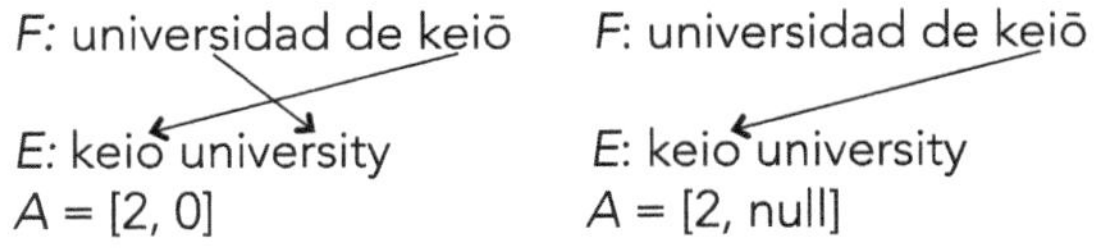

Fig. 5.8 Examples of the word alignment variable A in Eq. (5.2)

F: universidad de keiō
E: keio university
A = [2, 0]

F: universidad de keiō
E: keio university
A = [2, null]

They use the EM algorithm (Dempster et al. 1977) to maximize the likelihood of training pairs, and update the model parameters iteratively. At inference time, the translation of a foreign phrase F is generated according to Bayes rule:

$$E^* = \arg\max_E P(E|F) = \arg\max_E P(F|E)P(E)$$

where $P(F|E)$ is from Eq. (5.2) and $P(E)$ is obtained from an English language model. The translated name is then used to lookup an English dictionary for generating English candidate entities.

Pan et al. (2017) also use Wikipedia inter-language links to translate mentions into English for cross-lingual entity linking. For each word in a non-English mention, they mine its English translation using Wikipedia title pairs which contain this word. The most likely translation is extracted from these title pairs by GIZA++ (Och and Ney 2003). The translated mention is again used in a dictionary based approach for retrieving English candidate entities. This approach may work well for translating frequent words such as "Institute" or "Beijing" since they appear in many Wikipedia titles. But rare words especially for which require word transliteration may not be translated correctly.

5.3 Discussion and Conclusions

In this chapter, we discussed various candidate generation techniques for monolingual and cross-lingual entity linking problems. The goal of this step is to quickly produce a relatively small set of candidate entities a mention might refer to, so that models do not need to consider all entities in a knowledge base.

We classify candidate generation approaches into three categories: dictionary based, query log based, and retrieval based approaches. Dictionary based methods are most widely used due to its simplicity, especially when the target knowledge base is Wikipedia. Since Wikipedia contains plenty of structured information, such as anchor text, redirects, and disambiguation pages, which are very useful in constructing a dictionary that maps mentions to possible entities. Query log from search engines also has been shown useful for popular and large knowledge bases since there is a large number of users and looking up information in an encyclopedia becomes a norm nowadays. Dictionary based and query log based methods are usually used together. Better performance has been reported from this hybrid approach in several studies. However, for the smaller or less popular knowledge bases, these two approaches might not have a good coverage.

After 2019, more researchers proposed to use retrieval based approaches due to the thrive of deep learning and representation learning. This type of methods computes similarity between a mention and all entities in a knowledge base, usually based on some pre-computed embedding representations. It is less efficient than the dictionary based approach when the knowledge base is large. However, with proper indexing and searching techniques, fast retrieval is still possible. One advantage of retrieval based approaches is that it could work for knowledge bases which have less structured information, for example, when each entity only has a short paragraph of description. Since these methods leverage pre-trained large language models, the resulting entity representations could have a good quality even there is not much information in the knowledge base.

To apply monolingual candidate generation methods to the cross-lingual setting, the key challenge becomes how to reach entities in one language from mentions written in another language. Most works use Wikipedia or Wikidata as the target knowledge base due to its abundant information across languages. For example, the cross-language links in Wikipedia have been widely leveraged in the cross-lingual entity linking research. Another general approach we discussed in this chapter is to reduce the cross-lingual problem to monolingual problem via a translation or transliteration model. More recently, retrieval based methods with pre-trained multilingual large language models improve over dictionary based methods significantly.

5.4 Bibliographical Notes

Dictionary based candidate generation methods have been extremely popular in monolingual entity linking due to its efficency. Most works (Bunescu and Paşca 2006; Mihalcea and Csomai 2007; Cucerzan 2007; Kulkarni et al. 2009; Zhang et al. 2010; Zheng et al. 2010; Gottipati and Jiang 2011; Ratinov et al. 2011; Han and Sun 2011; Zhang et al. 2011; Shen et al. 2012; Guo et al. 2013; Gattani et al. 2013; Luo et al. 2015; Lazic et al. 2015; Globerson et al. 2016) build alias dictionaries from the English Wikipedia.

Besides using Wikipedia articles, Chisholm and Hachey (2015) also use hyperlinks on the web to collect more mention-to-entity mapping. Pershina et al. (2015) released the dictionary they built for AIDA-CoNLL dataset.[5] Although they did not provide details of how this candidate dictionary was built, the generated candidate sets have been show to have high recall and low ambiguity. This candidate generation resource has been used in several studies (Yamada et al. 2016; Sil et al. 2018; Yang et al. 2018).

In addition to Cucerzan (2007) that we discussed in this chapter, Gottipati and Jiang (2011), Pershina et al. (2015), Zhang et al. (2011), Zheng et al. (2010) also apply simple coreference resolution or surface form expansion to resolve the short mentions before generating candidate entities.

For the retrieval based approach for cross-lingual candidate generation, Zhou et al. (2020) further improved the cross-lingual entity similarity model proposed by Rijhwani et al. (2019). They incorporate more variation of names in training, such as partial names and alias from Wikidata. Also, they use bag of character n-grams to represent name strings instead of the original Bi-LSTM model on character sequence.

---

[5] https://github.com/masha-p/PPRforNED.

Earlier works that study name transliteration generation include Finch et al. (2015), Jiampojamarn et al. (2009, 2010), Li et al. (2004), Ravi and Knight (2009). Irvine et al. (2010), Virga and Khudanpur (2003) treat this problem as a machine translation problem, whereas Ammar et al. (2012), Pingali et al. (2008), Reddy and Waxmonsky (2009) view it as a sequence labeling problem.

References

Ammar, W., Dyer, C., Smith, N.: Transliteration by sequence labeling with lattice encodings and reranking. In: Proceedings of the 4th Named Entity Workshop (NEWS) 2012, Association for Computational Linguistics, pp. 66–70. Jeju, Korea (2012). https://aclanthology.org/W12-4410

Botha, J.A., Shan, Z., Gillick, D.: Entity Linking in 100 Languages. In: Proceedings of the 2020 Conference on Empirical Methods in Natural Language Processing (EMNLP), Association for Computational Linguistics, pp. 7833–7845. Online (2020). https://doi.org/10.18653/v1/2020.emnlp-main.630, https://aclanthology.org/2020.emnlp-main.630

Bunescu, R., Paşca, M.: Using encyclopedic knowledge for named entity disambiguation. In: 11th Conference of the European Chapter of the Association for Computational Linguistics, pp. 9–16. Trento, Italy (2006). https://aclanthology.org/E06-1002

Chang, M.-W., Goldwasser, D., Roth, D., Tu, Y.: Unsupervised constraint driven learning for transliteration discovery. In: Proceedings of Human Language Technologies: The 2009 Annual Conference of the North American Chapter of the Association for Computational Linguistics, Association for Computational Linguistics, pp. 299–307. Boulder, Colorado (2009). https://aclanthology.org/N09-1034

Chang, M.-W., Ratinov, L., Roth, D.: Structured learning with constrained conditional models. Machine Learn. **88**(3), 399–431 (2012). http://cogcomp.org/papers/ChangRaRo12.pdf

Chisholm, A., Hachey, B.: Entity disambiguation with web links. Trans. Assoc. Comput. Linguistics **3**, 145–156 (2015)

Cucerzan, S.: Large-scale named entity disambiguation based on Wikipedia data. In: Proceedings of the 2007 Joint Conference on Empirical Methods in Natural Language Processing and Computational Natural Language Learning (EMNLP-CoNLL), Association for Computational Linguistics, pp. 708–716. Prague, Czech Republic (2007). https://aclanthology.org/D07-1074

Darwish, K.: Named entity recognition using cross-lingual resources: arabic as an example. In: Proceedings of the 51st Annual Meeting of the Association for Computational Linguistics (Volume 1: Long Papers), Association for Computational Linguistics, pp. 1558–1567. Sofia, Bulgaria (2013). https://aclanthology.org/P13-1153

Dempster, A.P., Laird, N.M., Rubin, D.B.: Maximum likelihood from incomplete data via the EM algorithm. J. R. Stat. Soc., Series B (1977)

Dredze, M., McNamee, P., Rao, D., Gerber, A., Finin, T.: Entity disambiguation for knowledge base population. In: Proceedings of the 23rd International Conference on Computational Linguistics (Coling 2010), Coling 2010 Organizing Committee, pp. 277–285. Beijing, China (2010). https://aclanthology.org/C10-1032

Finch, A., Liu, L., Wang, X., Sumita, E.: Neural network transduction models in transliteration generation. In: Proceedings of the Fifth Named Entity Workshop, Association for Computational Linguistics, pp. 61–66. Beijing, China (2015). https://doi.org/10.18653/v1/W15-3909, https://aclanthology.org/W15-3909

Fu, X., Shi, W., Yu, X., Zhao, Z., Roth, D.: Design challenges in low-resource cross-lingual entity linking. In: Proceedings of the 2020 Conference on Empirical Methods in Natural Language Processing (EMNLP), Association for Computational Linguistics, pp. 6418–6432. Online (2020). https://doi.org/10.18653/v1/2020.emnlp-main.521, https://aclanthology.org/2020.emnlp-main.521

Ganea, O.-E., Ganea, M., Lucchi, A., Eickhoff, C., Hofmann, T.: Probabilistic bag-of-hyperlinks model for entity linking. In: Proceedings of the 25th International Conference on World Wide Web (WWW), pp. 927–938 (2016)

Gattani, A., Lamba, D.S., Garera, N., Tiwari, M., Chai, X., Das, S., Subramaniam, S., Rajaraman, A., Harinarayan, V., Doan, A.: Entity extraction, linking, classification, and tagging for social media: a Wikipedia-based approach. Very Large Data Base (VLDB) Endowment **6**(11), 1126–1137 (2013)

Gillick, D., Kulkarni, S., Lansing, L., Presta, A., Baldridge, J., Ie, E., Garcia-Olano. D.: Learning dense representations for entity retrieval. In: Proceedings of the 23rd Conference on Computational Natural Language Learning (CoNLL), Association for Computational Linguistics, pp. 528–537. Hong Kong, China (2019). https://doi.org/10.18653/v1/K19-1049, https://aclanthology.org/K19-1049

Globerson, A., Lazic, N., Chakrabarti, S., Subramanya, A., Ringgaard, M., Pereira, F.: Collective entity resolution with multi-focal attention. In: Proceedings of the 54th Annual Meeting of the Association for Computational Linguistics (Volume 1: Long Papers), Association for Computational Linguistics, pp. 621–631. Berlin, Germany (2016). https://doi.org/10.18653/v1/P16-1059, https://aclanthology.org/P16-1059

Gottipati, S., Jiang, J.: Linking entities to a knowledge base with query expansion. In: Proceedings of the 2011 Conference on Empirical Methods in Natural Language Processing, Association for Computational Linguistics, pp. 804–813. Edinburgh, Scotland, UK (2011). https://aclanthology.org/D11-1074

Guo, S., Chang, M.-W., Kiciman, E.: To link or not to link? a study on end-to-end tweet entity linking. In: Proceedings of the 2013 Conference of the North American Chapter of the Association for Computational Linguistics: Human Language Technologies, Association for Computational Linguistics, pp. 1020–1030. Atlanta, Georgia (2013). https://aclanthology.org/N13-1122

Gupta, N., Singh, S., Roth, D.: Entity linking via joint encoding of types, descriptions, and context. In: Proceedings of the 2017 Conference on Empirical Methods in Natural Language Processing, Association for Computational Linguistics, pp. 2681–2690. Copenhagen, Denmark (2017). https://doi.org/10.18653/v1/D17-1284, https://aclanthology.org/D17-1284

Hachey, B., Radford, W., Nothman, J., Honnibal, M., Curran, J.R.: Evaluating entity linking with wikipedia. Artif. intell. **194**, 130–150 (2013)

Han, X., Sun, L.: A generative entity-mention model for linking entities with knowledge base. In: Proceedings of the 49th Annual Meeting of the Association for Computational Linguistics: Human Language Technologies, Association for Computational Linguistics, pp. 945–954. Portland, Oregon, USA (2011). https://aclanthology.org/P11-1095

Irvine, A., Callison-Burch, C., Klementiev, A.: Transliterating from all languages. In: Proceedings of the 9th Conference of the Association for Machine Translation in the Americas: Research Papers, Association for Machine Translation in the Americas. Denver, Colorado, USA (2010). https://aclanthology.org/2010.amta-papers.12

Jaleel, N.A., Larkey, L.S.: Statistical transliteration for English-Arabic cross language information retrieval. In: Proceedings of the ACM Conference on Information and Knowledge Management (CIKM) (2003)

Jiampojamarn, S., Bhargava, A., Dou, Q., Dwyer, K., Kondrak, G.: DirecTL: a language independent approach to transliteration. In: Proceedings of the 2009 Named Entities Workshop: Shared Task on Transliteration (NEWS 2009), Association for Computational Linguistics, pp. 28–31. Suntec, Singapore (2009). https://aclanthology.org/W09-3504

Jiampojamarn, S., Dwyer, K., Bergsma, S., Bhargava, A., Dou, Q., Kim, M.-Y., Kondrak, G.: Transliteration generation and mining with limited training resources. In: Proceedings of the 2010 Named Entities Workshop, Association for Computational Linguistics, pp. 39–47. Uppsala, Sweden (2010). https://aclanthology.org/W10-2405

Johnson, J., Douze, M., Jégou, H.: Billion-scale similarity search with GPUs. IEEE Trans. Big Data (2019)

Karimi, S., Scholer, F., Turpin, A.: Machine transliteration survey. ACM Comput. Surv. (2011). https://doi.org/10.1145/1922649.1922654

Klementiev, A., Roth, D.: Named entity transliteration and discovery in multilingual corpora. In: Goutte, C., Cancedda, N., Dymetman, M., Foster, G. (eds.) Learning Machine Translation. MIT Press (2008). http://cogcomp.org/papers/KlementievRo08.pdf

Knight, K., Graehl, J.: Machine transliteration. In: 35th Annual Meeting of the Association for Computational Linguistics and 8th Conference of the European Chapter of the Association for Computational Linguistics, Association for Computational Linguistics, pp. 128–135. Madrid, Spain (1997). https://doi.org/10.3115/976909.979634, https://aclanthology.org/P97-1017

Koehn, P., Hoang, H., Birch, A., Callison-Burch, C., Federico, M., Bertoldi, N., Cowan, B., Shen, W., Moran, C., Zens, R., Dyer, C., Bojar, O., Constantin, A., Herbst, E.: Moses: open source toolkit for statistical machine translation. In: Proceedings of the 45th Annual Meeting of the Association for Computational Linguistics Companion Volume Proceedings of the Demo and Poster Sessions, Association for Computational Linguistics, pp. 177–180. Prague, Czech Republic (2007). https://aclanthology.org/P07-2045

Kulkarni, S., Singh, A., Ramakrishnan, G., Chakrabarti, S.: Collective annotation of Wikipedia entities in web text. In: Proceedings of the 15th ACM SIGKDD Conference on Knowledge Discovery and Data Mining (KDD), pp. 457–466. ACM (2009)

Lazic, N., Subramanya, A., Ringgaard, M., Pereira, F.: Plato: a selective context model for entity resolution. Trans. Assoc. Comput. Linguistics **3**, 503–515 (2015)

Li, Z., Callison-Burch, C., Dyer, C., Khudanpur, S., Schwartz, L., Thornton, W., Weese, J., Zaidan, O.: Joshua: an open source toolkit for parsing-based machine translation. In: Proceedings of the Fourth Workshop on Statistical Machine Translation, Association for Computational Linguistics, pp. 135–139. Athens, Greece (2009). https://aclanthology.org/W09-0424

Li, H., Zhang, H., Su, J.: A joint source-channel model for machine transliteration. In: Proceedings of the 42nd Annual Meeting of the Association for Computational Linguistics (ACL-04), pp. 159–166. Barcelona, Spain (2004). https://doi.org/10.3115/1218955.1218976, https://aclanthology.org/P04-1021

Ling, X., Singh, S., Weld, D.S.: Design challenges for entity linking. Trans. Assoc. Comput. Linguistics **3**, 315–328 (2015)

Logeswaran, L., Chang, M.-W., Lee, K., Toutanova, K., Devlin, J., Lee, H.: Zero-shot entity linking by reading entity descriptions. In: Proceedings of the 57th Annual Meeting of the Association for Computational Linguistics, Association for Computational Linguistics, pp. 3449–3460. Florence, Italy (2019). https://doi.org/10.18653/v1/P19-1335, https://aclanthology.org/P19-1335

Luo, G., Huang, X., Lin, C.-Y., Nie, Z.: Joint entity recognition and disambiguation. In: Proceedings of the 2015 Conference on Empirical Methods in Natural Language Processing, Association for Computational Linguistics, pp. 879–888. Lisbon, Portugal (2015). https://doi.org/10.18653/v1/D15-1104, https://aclanthology.org/D15-1104

Mihalcea, R., Csomai, A.: Wikify!: linking documents to encyclopedic knowledge. In: Proceedings of the ACM Conference on Information and Knowledge Management (CIKM) (2007)

Och, F.J., Ney, H.: A systematic comparison of various statistical alignment models. Comput. Linguistics **29**(1), 19–51 (2003)

Pan, X., Zhang, B., May, J., Nothman, J., Knight, K., Ji, H.: Cross-lingual name tagging and linking for 282 languages. In: Proceedings of the 55th Annual Meeting of the Association for Computational Linguistics (Volume 1: Long Papers), Association for Computational Linguistics, pp. 1946–1958. Vancouver, Canada (2017). https://doi.org/10.18653/v1/P17-1178, https://aclanthology.org/P17-1178

Pasternack, J., Roth, D.: Learning better transliterations. In: Proceedings of the ACM Conference on Information and Knowledge Management (CIKM), vol. 11 (2009). http://cogcomp.org/papers/PasternackRo09a.pdf

Pershina, M., He, Y., Grishman, R.: Personalized page rank for named entity disambiguation. In: Proceedings of the 2015 Conference of the North American Chapter of the Association for Computational Linguistics: Human Language Technologies, Association for Computational Linguistics, pp. 238–243. Denver, Colorado (2015). https://doi.org/10.3115/v1/N15-1026, https://aclanthology.org/N15-1026

Pingali, P., Ganesh, S., Yella, S., Varma, V.: Statistical transliteration for cross language information retrieval using HMM alignment model and CRF. In: Proceedings of the 2nd workshop on Cross Lingual Information Access (CLIA) Addressing the Information Need of Multilingual Societies (2008). https://aclanthology.org/I08-6006

Ratinov, L., Roth, D., Downey, D., Anderson, M.: Local and global algorithms for disambiguation to Wikipedia. In: Proceedings of the 49th Annual Meeting of the Association for Computational Linguistics: Human Language Technologies, Association for Computational Linguistics, pp. 1375–1384. Portland, Oregon, USA (2011). https://aclanthology.org/P11-1138

Ravi, S., Knight, K.: Learning phoneme mappings for transliteration without parallel data. In: Proceedings of Human Language Technologies: The 2009 Annual Conference of the North American Chapter of the Association for Computational Linguistics, pp. 37–45. Boulder, Colorado (2009). https://aclanthology.org/N09-1005

Reddym, S., Waxmonsky, S.: Substring-based transliteration with conditional random fields. In: Proceedings of the 2009 Named Entities Workshop: Shared Task on Transliteration (NEWS 2009), Association for Computational Linguistics, pp. 92–95. Suntec, Singapore (2009). https://aclanthology.org/W09-3520

Rijhwani, S., Xie, J., Neubig, G., Carbonell, J.: Zero-shot neural transfer for cross-lingual entity linking. In: Proceedings of the AAAI Conference on Artificial Intelligence, vol. 33, pp. 6924–6931 (2019)

Robertson, S.E., Walker, S.: Some simple effective approximations to the 2-poisson model for probabilistic weighted retrieval. In: Proceedings of the Seventeenth Annual International ACM-SIGIR Conference on Research and Development in Information Retrieval, pp. 232–241. Springer (1994)

Shen, W., Wang, J., Luo, P., Wang, M.: LINDEN: linking named entities with knowledge base via semantic knowledge. In: Proceedings of the 21st international conference on World Wide Web (WWW), pp. 449–458. ACM (2012)

Sil, A., Kundu, G., Florian, R., Hamza, W.: Neural cross-lingual entity linking. In:0 Proceedings of the Conference on Artificial Intelligence (AAAI) (2018)

Sil. A., Florian, R.: One for all: towards language independent named entity linking. In: Proceedings of the 54th Annual Meeting of the Association for Computational Linguistics (Volume 1: Long Papers), Association for Computational Linguistics, pp. 2255–2264. Berlin, Germany (2016). https://doi.org/10.18653/v1/P16-1213, https://aclanthology.org/P16-1213

Spitkovsky, V.I., Chang, A.X.: A cross-lingual dictionary for English Wikipedia concepts. In: Proceedings of the Eighth International Conference on Language Resources and Evaluation (LREC'12), European Language Resources Association (ELRA), pp. 3168–3175. Istanbul, Turkey (2012). http://www.lrec-conf.org/proceedings/lrec2012/pdf/266_Paper.pdf

Sproat, R., Tao, T., Zhai, C.: Named entity transliteration with comparable corpora. In: Proceedings of the 21st International Conference on Computational Linguistics and 44th Annual Meeting of the Association for Computational Linguistics, pp. 73–80. Sydney, Australia (2006). https://doi.org/10.3115/1220175.1220185, https://aclanthology.org/P06-1010

Tsai, C.-T., Roth, D.: Concept grounding to multiple knowledge bases via indirect supervision. Trans. Assoc. Comput. Linguistics **4**, 141–154 (2016)

Tsai, C.-T., Roth, D.: Cross-lingual wikification using multilingual embeddings. In: Proceedings of the 2016 Conference of the North American Chapter of the Association for Computational Linguistics: Human Language Technologies, Association for Computational Linguistics, pp. 589–598. San Diego, California (2016b). https://doi.org/10.18653/v1/N16-1072, https://aclanthology.org/N16-1072

Tsai, C.-T., Roth, D.: Learning better name translation for cross-lingual Wikification. In: Proceedings of the Conference on Artificial Intelligence (AAAI), vol. 2 (2018). http://cogcomp.org/papers/TsaiRo18.pdf

Virga, P., Khudanpur, S.: Transliteration of proper names in cross-lingual information retrieval. In: Proceedings of the ACL 2003 Workshop on Multilingual and Mixed-language Named Entity Recognition, Association for Computational Linguistics, pp. 57–64. Sapporo, Japan (2003). https://doi.org/10.3115/1119384.1119392, https://aclanthology.org/W03-1508

Wu, L., Petroni, F., Josifoski, M., Riedel, S., Zettlemoyer, L.: Scalable zero-shot entity linking with dense entity retrieval. In: Proceedings of the 2020 Conference on Empirical Methods in Natural Language Processing (EMNLP), Association for Computational Linguistics, pp. 6397–6407. Online (2020). https://doi.org/10.18653/v1/2020.emnlp-main.519, https://aclanthology.org/2020.emnlp-main.519

Yamada, I., Shindo, H., Takeda, H., Takefuji, Y.: Joint learning of the embedding of words and entities for named entity disambiguation. In: Proceedings of the 20th SIGNLL Conference on Computational Natural Language Learning, Association for Computational Linguistics, pp. 250–259. Berlin, Germany (2016). https://doi.org/10.18653/v1/K16-1025, https://aclanthology.org/K16-1025

Yang, Y., Irsoy, O., Rahman, K.S.: Collective entity disambiguation with structured gradient tree boosting. In: Proceedings of the 2018 Conference of the North American Chapter of the Association for Computational Linguistics: Human Language Technologies, Volume 1 (Long Papers), Association for Computational Linguistics, pp. 777–786. New Orleans, Louisiana (2018). https://doi.org/10.18653/v1/N18-1071, https://aclanthology.org/N18-1071

Zhang, W., Sim, Y.-C., Su, J., Tan, C.-L.: Entity linking with effective acronym expansion, instance selection and topic modeling. In: Twenty-Second International Joint Conference on Artificial Intelligence (IJCAI) (2011)

Zhang, W., Su, J., Tan, C.L., Wang, W.T.: Entity linking leveraging automatically generated annotation. In: Proceedings of the 23rd International Conference on Computational Linguistics (Coling 2010), Coling 2010 Organizing Committee, pp. 1290–1298. Beijing, China (2010). https://aclanthology.org/C10-1145

Zheng, Z., Li, F., Huang, M., Zhu, X.: Learning to link entities with knowledge base. In: Human Language Technologies: The 2010 Annual Conference of the North American Chapter of the Association for Computational Linguistics, Association for Computational Linguistics, pp. 483–491. Los Angeles, California (2010). https://aclanthology.org/N10-1072

Zhou, S., Rijhwani, S., Wieting, J., Carbonell, J., Neubig, G.: Improving candidate generation for low-resource cross-lingual entity linking. Trans. Assoc. Comput. Linguistics **8**, 109–124 (2020)

Linking Mentions to Entities

In the previous two chapters, we have discussed the first two components of the entity linking pipeline: locating mentions and generating candidate entities. The next step in the pipeline is to select an entity from the candidate set of each mention according to the meaning of the input text. In the candidate generation step, we see that most approaches are solely based on name string similarity and entity popularity, namely, other contextual information from the input text is not used. Since the goal of candidate generation is to quickly produce a small set of entities that contains the target entity, sophisticated models might not be needed. However, in order to correctly pick the answers out of the candidate sets, contextual information should be taken into account. This step is often referred as context-sensitive inference.

Figure 6.1 shows an example of inputs and outputs of this step. This context-sensitive inference step is usually viewed as a ranking problem, where a mention is a query and the corresponding candidate entities are the possible outcomes. A model in this step will assign a score to each candidate, which indicates how relevant it is to the mention in the context. The entity with the highest relevance score will then be returned as the answer.

The cross-lingual setup is conceptually very similar to the monolingual setup, just the mentions are written in one language but candidate entities are in another language's knowledge base. If one can measure textual similarity across two languages, as we have discussed in the previous two chapters (cross-lingual NER and candidate generation), the inference models developed for English could potentially be applied to the cross-lingual case directly.

Fig. 6.1 An example of input and output of the context sensitive inference step. The input consists of a list of entity candidates for each mention. A model in this step assigns a score to each entity based on how relevant it is to the context. The candidates with the highest score of each mention will be selected as the answers

6.1 Formulating the Linking Problem

We start by formulating the linking problem formally. Let $m_1, \ldots, m_n$ be the mentions identified in a given piece of text, and $C(m_i)$ be the set of candidate entities of m_i retrieved by a candidate generation algorithm. The objective of the linking problem is essentially to pick one candidate entity for each mention so that some scoring function Ω is maximized:

$$\hat{e}_1, \ldots, \hat{e}_n = \underset{e_i \in C(m_i) \forall i \in \{1, \ldots, n\}}{\arg \max} \Omega(m_1 \to e_1, \ldots, m_n \to e_n). \quad (6.1)$$

We can see that this optimization problem becomes intractable quickly as the number of mentions grows, if all combinations of candidate entities need to be considered. For example, if there are 5 mentions and each mention has 20 candidate entities, there are $20^5 = 3.2$ million possible answers. A naive solution to this problem is to link each mention individually. Using this approach, we may lose some valuable information since the decision on one mention is independent of other mentions' assignments. Figure 6.2 shows an example

Fig. 6.2 When making decision for multiple mentions simultaneously, the relationship between candidate entities could provide useful signals. In contrast, if we do local inference only on "Socialist Party", it might be hard to link to the correct entity `Socialist_Party_of_Serbia` since it is not a very popular entity

of a potential benefit of resolving two mentions simultaneously. From this example, we see that the relationships between entities in a KB could strongly indicate that the two entities are likely to co-occur. Therefore, there is a trade-off between efficiency and accuracy when solving Eq. 6.1. Researchers have been proposed various ways and made different assumptions to overcome this scaling challenge. In Sect. 6.3, we will discuss and categorize approaches that solve this inference problem.

Another key issue is that what information goes into the scoring function Ω, which measures the relevancy between mentions and entities. What kinds of features could characterize the compatibility between a mention and an entity, and what features could capture the relationships between two mentions or between two entities? The next section will be devoted to discussing features proposed in the literature.

6.2 Features

Features are properties of mentions and entities designed for capturing how likely an entity is the answer for a mention. Features can be hand-crafted as in more traditional machine learning approaches or learnt by neural network models. Supervised methods usually use multiple features and learn a model to combine these features, whereas unsupervised methods may only consider a couple of features.

We categorize features according to what kind of information is used. At the top level, we broadly classify features into *local* and *global* features. Local features are based on single mention and the corresponding candidate entities, whereas global features use other mentions in the input text along with their candidate entities. Figure 6.3 illustrates local features ϕ and global features ψ. For example, the local feature $\phi(m_1 \to e_1)$ only uses context of m_1 (Milošević) and/or knowledge base information of e_1 (Slobodan_Milošević). In contrast, the global feature $\psi(m_1 \to e_1, m_2 \to e_4)$ will look at the relationships between m_1 and m_2, or the connections between e_1 and e_4 in the KB. Note that global features may introduce inter-dependency between mentions, which could make the inference problem

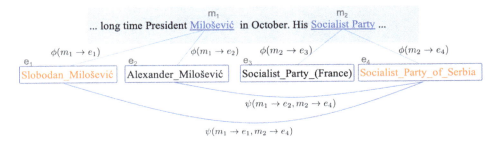

Fig. 6.3 Illustration of local features ϕ and global features ψ

much harder. We will discuss more about this issue in the next section when we introduce inference models.

Both local and global features can be further categorized into *knowledge base features* and *document context features*. Knowledge base features usually do not look at words other than the target mention in the input document. For instance, one popular global feature is whether two entities are linked by some relation in the KB. Due to this context-independent nature, local features of this type are usually used for the initial ranking in the candidate generation step. On the other hand, document context features measure similarity between mentions and candidate entities using other words or phrases in the given documents. The fundamental difference between document context features is how mentions are represented by context in the documents and how entities are represented by information in the knowledge base, so that some similarity between mention and entity representations can lead to the target entity.

6.2.1 Local Features

All features introduced in this section are of the form $\phi_i(m \to e)$ as illustrated in Fig. 6.3. These features only consider a single mention-entity pair.

6.2.1.1 Knowledge Base Features

This type of features does not depend on the context of input text. Instead, they only use the mention surface string and information in the knowledge base. Many of these features are also used in the candidate generation step to produce an initial ranking of candidate entities.

- **Entity popularity**: Almost all entity linking systems use some entity popularity features since well-known entities are more likely to be mentioned. When using Wikipedia as the target knowledge base, the most common entity popularity feature is the probability of the candidate entity e given the mention m, the prior probability that is used in the candidate generation step:

$$\Pr(e|m) = \frac{\text{\# times } m \text{ is linked to } e}{\text{\# times } m \text{ is hyperlinked}}. \tag{6.2}$$

This probability can be estimated from the anchor texts in Wikipedia or documents on the web. Besides this prior, Dredze et al. (2010) use features derived from the Wikipedia graph structure to represent entity popularity, including indegree of a page, outdegree of a page, and Wikipedia page length in bytes. In addition, Guo et al. (2013) use view statistics of Wikipedia pages to measure entity popularity:

$$\Pr(e|m) = \frac{\text{view count of } e}{\sum_{i=1}^{|C(m)|} \text{view count of } e_i}.$$

Instead of using the raw view count of entity e, they normalize the raw view count by the total view counts of entities in the candidate set $C(m)$.
- **Name similarity**: This type of features compute string similarity between mentions and entity names in the knowledge base. Some features are based on string similarity measures such as edit distance, character Dice, or Hamming distance. Other name comparison features that have been used in the literature include: whether the entity name matches the mention exactly, whether the entity name starts or ends with the mention, and the number of identical words between the entity name and the mention. In addition, a knowledge base usually contains a list of alias for each entity, which is leveraged to handle nicknames (e.g., New York City and Big Apple) and acronyms (e.g., New York City and NYC).

6.2.1.2 Document Context Features
- **Entity type**: The match of entity types between mentions and candidate entities has been shown to be useful for entity linking. Since a named entity recognition model is usually applied to extract mentions, each mention will also be assigned an entity type (e.g., person, location, organization).[1] Although these are coarse-grained types, they could be useful for distinguishing eponymous entities. For instance, if "John F. Kennedy International Airport" is tagged as a location, candidate entities that refer to people should be less likely to be the answer. A challenge of this approach is how to get the NER entity types for entries in a knowledge base. For Wikipedia articles, Dredze et al. (2010) infer entity types from the information in the infobox. Another common way is to get typing information from the corresponding entries in Freebase (Ling et al. 2015), since the typing taxonomy used in Freebase is closer to the common NER types.
- **Contextual similarity**: In this category, mentions are represented by the words in the input text. For instance, the context of a mention could be the words in the entire document, words in a small window around the mention, or other mentions in the document. For an entity in Wikipedia, the context could be the words in the entire Wikipedia page, words in the first paragraph, words around the anchor texts which point to the entity page, anchor texts in the entity page, or words in the Wikipedia categories of the page. Besides the simplest bag-of-words representations, one could also weight each word by the TF-IDF scores (Ratinov et al. 2011). After mention and entity representations are constructed, several similarity measures have been applied in the literature, including cosine similarity, dot product, word overlap, and Jaccard similarity.

Neural network models have become popular in recent years. Several models are proposed to generate distributional representations for mentions and entities for the entity linking task. In the following paragraphs, we briefly discuss representative works that use neural network models to learn features.

---
[1] See Chap. 4 for more details of NER.

The most straightforward way is to represent words with some pre-trained word embeddings (e.g., skip-gram model), and then combine context words of a mention or an entity into a single vector. Approaches for this combination step could be, for example, averaging (Yamada et al. 2016) or through convolutional neural networks (Francis-Landau et al. 2016).

Instead of combining word representations into entity representations, several models have been proposed to generate entity embeddings for entity linking task. He et al. (2013) jointly optimize document and entity representations for a given similarity measure. They apply stacked denoising auto-encoders to generate document representations, and the anchor texts in Wikipedia are then used as supervision to fine tune both document and entity representations toward the similarity measure. Yamada et al. (2016) learn word and entity representations jointly by applying skip-gram models on Wikipedia documents and a knowledge graph. By substituting anchor texts with the corresponding entities, words and entities are mapped into the same continuous vector space. Gupta et al. (2017) proposed a neural network model to encode each entity in Wikipedia. The model explicitly uses multiple sources of information of each entity, such as its description, contexts around its hyperlinked mentions, and its fine-grained types from Freebase.

Logeswaran et al. (2019) represent mentions in context and entity descriptions by BERT model.[2] They encode each pair of mention with context and entity description using a Transformer model. That is, instead of learning a representation for each mention and entity, the proposed model learns a representation for each mention-entity pair, so that cross-attention between mentions and entities can be explore by a deep architecture. Several recent works follow a similar idea (e.g., Wu et al. (2020), Li et al. (2020a)).

6.2.1.3 Cross-Lingual Variants

Several aforementioned features can be used directly in the cross-lingual entity linking setup. For example, entity popularity measures such as view counts and indegree and outdegree of an entity do not depend on the input text, so these features are identical in the monolingual and cross-lingual settings. However, the entity prior defined in Eq. (6.2) does depend on the mention, which is written in a different language in the cross-lingual setup. If the target knowledge base is Wikipedia, we can still compute this prior with the help of the inter-language links.[3] That is, we can count the number of times that a mention in language A is linked to an entity in language B by going through the corresponding entity in the language B's Wikipedia first. Similar to what we have discussed in the candidate generation chapter, the quality of this cross-lingual prior largely depends on the size of language B's Wikipedia, since the target entity has to be present in the language B's Wikipedia and there has to be an inter-language link between this entity's pages in both Wikipedia.

---

[2] BERT is introduced in Appendix A.2.2.
[3] Inter-language links in Wikipedia were discussed in Sect. 2.1.1.

6.2 Features

Most cross-lingual entity linking work focuses on the contextual similarity features. The problem is essentially cross-lingual similarity between mentions in one language and entities in another language. Naturally, cross-lingual word embedding models serve as the key building block of this type of features. We will discuss a few cross-lingual word embeddings in Appendix A.1.4. If mentions and entities are both represented by words, any pre-trained cross-lingual word embedding can be used to encode them in theory. Nevertheless, several work proposed to train better entity embeddings specifically for the cross-lingual entity linking setup.

For instance, Tsai and Roth (2016b) proposed to jointly train cross-lingual embeddings for words and Wikipedia titles by the skip-gram objective. The proposed method leverages anchor texts and inter-language links in multilingual Wikipedia, therefore it can be applied to all languages represented in Wikipedia. These pre-trained representations of words and Wikipedia titles are then used to compute some local and global similarity features for a ranking model. Upadhyay et al. (2018) use FastText (Smith et al. 2017), a multilingual word embedding model, to encode mentions in any supported language. They proposed a model to learn entity representations through a cross-lingual entity linking objective. Figure 6.4 shows diagrams of their proposed mention context encoder (Fig. 6.4a) and the loss between different source of information (Fig. 6.4b). Similar to the trend we have seen in

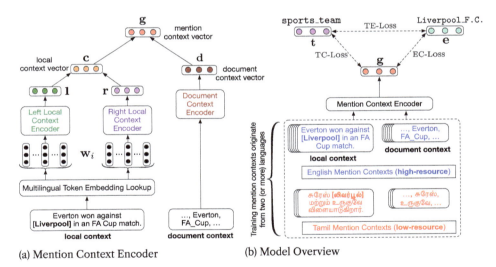

Fig. 6.4 The cross-lingual entity linking model proposed by Upadhyay et al. (2018). **a** The Mention Context Encoder encodes the local context (neighboring words) and the document context (surfaces of other mentions in the document) of the mention into **g**. **b** The model uses grounded mentions from two or more languages (English and Tamil shown) as supervision. The context **g**, entity **e** and type **t** vectors interact through Entity-Context loss (EC- Loss), Type-Context loss (TC- Loss) and Type-Entity loss (TE- Loss)

Chapter 5 and most of other NLP applications, pre-trained multilingual large language models are leveraged to solve cross-lingual entity linking task (Botha et al. 2020).

6.2.2 Global Features

Local features focus on the similarity or relevancy between a mention and its candidate entities. The global features that we are going to introduce in this section aim at capturing relationship between entities and between mentions in the input document. The idea is that the correct candidate should be more coherent or more related to other entities in the document than other candidates are. Namely, global features try to capture topical coherency between entities. All features introduced in this section can be expressed in the form of $\psi(m_i \rightarrow e_k, m_j \rightarrow e_l)$. These features could use information from two mentions and their candidate entities.

The key challenge of using entity coherence features is that decisions for different mentions become interdependent. Generating features for one mention depends on the disambiguation results of other mentions, and vice versa. When the number of mentions grows, the inference problem could beocome intractable quickly. We will discuss how researchers address this issue in the next section when we talk about inference models.

6.2.2.1 Knowledge Base Features
This category of features measure similarity between candidate entities of different mentions. When Wikipedia is the target knowledge base, researchers have proposed several kinds of similarity measures for two Wikipedia pages:

- Overlap between words in the two page.
- Overlap between Wikipedia categories or FreeBase types.
- Overlap between extracted key-phrases in the two pages.
- Number of links between the two pages.
- Overlap between incoming and outgoing links of the page.
- Similarity between the two entity embeddings.

Besides simply counting the number of overlaps, any similarity metric can be applied on these entity representations. In the literature, researchers have used, for instance, Normalized Google Distance, Point-wise Mutual Information, and Jaccard similarity.

6.2.2.2 Document Context Features
Global knowledge base features capture relationship or similarity between two entities according to their representations. Textual clues between the two corresponding mentions could also provide useful signals. Figure 6.5 illustrates the idea. If a coreference resolution

6.3 Inference Models

Fig. 6.5 An example of how textual relations could provide useful signals for disambiguating mentions. This example is adopted from Cheng and Roth (2013)

model recognizes that "Slobodan Milošević" and "Milošević" are coreferent, their entity linking results should match. Moreover, since mentions "Milošević" and "Socialist Party" have possessive relation, the candidate pairs which have a `founder_of` relation should be more likely to be the correct answer.

An example work of using this type of information is Cheng and Roth (2013). They extract syntactico-semantic relations and coreference relations among mentions. These relations are not only used to enrich candidate sets, but also to generate a relational score for each pair of candidate entities. This relational score can be viewed as a feature that captures some coherency between the two entities, and is used in their proposed relational inference objective.

6.2.2.3 Cross-Lingual Variants

Most cross-lingual entity linking work only uses local features, since the main focus is to produce better cross-lingual similarity between mentions and entities. A couple of models do try to incorporate some global knowledge base features. For instance, Tsai and Roth (2016b) measure the coherence between two entities by the cosine similarity of the proposed Wikipedia title embeddings.

Using global document context features in the cross-lingual scenario would be much more difficult since a relation extraction or coreference resolution model is needed for a non-English language. These tasks in themselves are pretty challenging especially for low-resource languages.

6.3 Inference Models

An inference model is responsible for selecting the best candidate entity for each mention using the features discussed in the previous section. There are two types of inference strategies: *local inference* and *global inference*. Local inference models make decision on each mention independently, whereas global inference models select the best entity for multiple mentions simultaneously. As we have seen in Fig. 6.2, a global inference model may result in more coherent assignments since the answer of one mention would affect the outcomes of the other mentions. However, considering all combinations of candidate entities may make global inference computational infeasible.

In this section, we will formulate the problem of local and global inference, and discuss some approximations of global inference which are not as expensive to compute but can incorporate some global information. We note that these inference models can be applied on both monolingual and cross-lingual settings.

6.3.1 Local Inference

Local inference resolves each mention m in a document $\mathcal{D}$ in isolation, with the assignment to mention m_i having no influence to assignment to mention m_j where $i \neq j$. To find the best entity $\hat{e}_i$ for mention m_i, we solve

$$\hat{e}_i = \arg\max_{e \in C(m_i)} \Phi(m_i \to e) \quad \forall m_i \in \mathcal{D}, \tag{6.3}$$

where $C(m_i)$ is the set of candidate entities of m_i, and $\Phi(m_i \to e)$ is a local scoring function which uses local features ϕ_k extracted from m_i and e. Namely, each mention m_i in the document is linked to entity $\hat{e}_i$ which has the highest score among all candidate entities. Local inference is a popular approach owing to its simplicity, and it is usually sufficient for resolving most common entities.

6.3.2 Global Inference

The local inference approach does not take into account relationship between candidate entities of different mentions in the same document. For instance, if Steven_Gerrard is a candidate for mention $m_i \in \mathcal{D}$ and Liverpool_F.C. is a candidate for mention $m_j \in \mathcal{D}$, then assigning $m_i \to$ Steven_Gerrard, $m_j \to$ Liverpool_F.C. is *coherent* because these two entities are related.[4] This correlation between entities is expressed through a pairwise coherence score $\Psi(m_i \to e_i, m_j \to e_j)$ that is used along with the local scoring function $\Phi(m_i \to e_i)$ to formulate a global inference problem,

$$(\hat{e}_1, \hat{e}_2, \ldots, \hat{e}_n) = \arg\max_{\substack{(e_1,\ldots,e_n) \in \\ C(m_i) \times \cdots \times C(m_n)}} \sum_{i=1}^{n} \Phi(m_i \to e_i) + \sum_{i \neq j} \Psi(m_i \to e_i, m_j \to e_j), \tag{6.4}$$

---

[4] Steven_Gerrard captained Liverpool_F.C. from 2003-15.

6.3 Inference Models

where n is the number of mentions in the input document. The pairwise coherence score Ψ can take global features that we discussed in Sect. 6.2.2. This pairwise score measures compatibility between two entities in the document and is usually based on relationships between entities. Since mentions are interdependent in this formulation, the model has to consider all combinations of candidate entities and make a joint decision.

This global inference problem is NP-hard (Cucerzan 2007), as the number of entity combinations grows exponentially with the number of mentions. Therefore, dependency assumption among mentions or approximate inference approaches that decompose the inference problem to smaller but tractable inference problems are often needed. For instance, Guo et al. (2013) assume linear structure dependency between mentions. They order mentions from left to right in the document and apply beam search algorithm to find the joint assignment approximately. In Ganea et al. (2016), Globerson et al. (2016), Ganea and Hofmann (2017), they all apply approximate inference techniques based on message passing to perform global inference.[5] Cheng and Roth (2013) use integer linear programming (ILP) to jointly select entities for multiple mentions. This ILP formulation utilizes both local ranking scores and relational scores from pairs of entities. We discuss this approach and introduce ILP formalism in Appendix A.4.1.

Another common approach is to reduce the inference problem in Eq. (6.4) to

$$\hat{e}_i = \arg\max_{e \in C(m_i)} \Phi(m_i \to e) + \sum_{e_k \in G(m_i)} \varphi(e, e_k), \qquad (6.5)$$

where $G(m_i)$ is a set of entities which are likely to present in the document, and $\varphi(e, e_k)$ is a scoring function measures similarity between two entities. We can see that this inference problem becomes local, that is, there is no dependency between mentions and each mention is resolved in isolation. Nevertheless, the second term in equation (6.5) still provides some entity coherence scores which could help to disambiguate m_i. How well does this formulation approximate global inference depends on how well this global entity set $G(m_i)$ represents the true entities exist in the document.

The inference models that use Eq. (6.5) usually consist of two steps. The first step is to find $G(m_i)$ for each mention, and then the second step uses this $G(m_i)$ to solve the inference problem. For example, Ratinov et al. (2011) and Yang and Chang (2015) both apply a local model in the first step to disambiguate each mention individually. This local model does not use any entity coherence features. The top k ranked entities of other mentions in the document are then collected to form $G(m_i)$. After obtaining $G(m_i)$, some global features based on entity pairs (Sect. 6.2.2) can be added to the model for re-training. For another example, Milne and Witten (2008) simply use the entities from all unambiguous mentions in the document as $G(m_i)$. This approach relies on the presence of unambiguous mentions with high disambiguation utility.

---

[5] We introduce belief propagation algorithm in Appendix A.4.2.

6.4 Learning Scoring Functions

The remaining question is what are those scoring functions $\Phi(m \to e)$ and $\Psi(m_i \to e_i, m_j \to e_j)$, which take features as inputs and produce a score. These functions are usually learnt by a machine learning model. We broadly categorize entity linking models into two groups based on the learning paradigms: supervised methods and unsupervised methods.

6.4.1 Supervised Methods

Supervised methods learn how to score candidates using a training corpus which contains labeled examples. Labeled examples for entity linking are basically mention-entity pairs, which could guide the models to better combine various features. Supervised methods are usually considered expensive in NLP applications since human annotators need to be trained to understand the task, and doing manual annotation is time consuming. However, for entity linking with Wikipedia, users of Wikipedia have manually created tons of labeled examples (anchor texts) in Wikipedia articles. Even in the cross-lingual case, cross-lingual supervision can be automatically generated by leveraging the inter-language links as we discussed in Chap. 3. Therefore, many entity linking systems leverage this free resource and apply supervised models. Nevertheless, we note that Wikipedia articles may not be the most ideal supervision for all kinds of models and text genres, since the format of Wikipedia articles are very formal, many anchor texts are not named entities, and many linkable mentions are not linked (only the first mention of an entity is linked in a Wikipedia article).

Most supervised models use various features for each mention-entity pair, each of which measures some aspect of relevancy between a mention and a candidate entity. We group supervised models into four categories: binary classification, learning to rank, structured prediction, and generative model. Different models make different assumptions and optimize different objectives.

6.4.1.1 Binary Classification

Binary classification models view each (mention, candidate entity) pair as a binary decision problem. Namely, weather the mention refers to the entity or not. At training time, the correct entities are labeled as positive instances. Other candidates generated by a candidate generation module are treated as negative instances. Some binary classifiers that have been applied to the entity linking problem are support vector machines (SVMs) (Milne and Witten 2008; Pilz and Paaß 2011; Zhang et al. 2010), logistic regression (Sil and Yates 2013), naïve Bayes (Milne and Witten 2008), and decision trees (Milne and Witten 2008).

One issue of binary classification formulation is that more than one candidate entities could be classified as positive at inference time, since a classifier makes decision for each mention independently. Milne and Witten (2008) do not resolve this issue and simply link

6.4 Learning Scoring Functions

a mention with multiple entities if there are more than one positive candidates. To address this issue (Pilz and Paaß 2011) pick the candidate which has the highest decision value from an SVM model. Zhang et al. (2010) use a vector space model which is based on several features used in the binary classifier to break ties.

6.4.1.2 Learning to Rank

In contrast to binary classification methods which make decision on each candidate independently of other candidates, learning to rank approaches model the preferences between candidate entities. Therefore, ranking models usually perform much better than binary classification models (Zheng et al. 2010). At prediction time, a ranking model assigns a score to each candidate, and the candidates are ranked according to these scores. The candidate with the highest score will be chosen as the answer.

When training a ranking model, an ordered list of outcomes is usually provided. Most ranking models perform pairwise comparison among candidates. However, training examples for entity linking problem are (mention, title) pairs. Instead of knowing the preferences of all pairs of candidates, the models only learn from knowing that the correct entity is more preferable than all the other entities in the candidate set. The model will not compare all pairs of candidates. Instead, it only uses the pairs with known preferences.

Most earlier entity linking systems that apply learning to rank framework (Bunescu and Paşca 2006; Ratinov et al. 2011; Tsai and Roth 2016b) use the RankSVM model, which is a pairwise ranking approach. Zheng et al. (2010), Chen and Ji (2011) also experiment with ListNet (Cao et al. 2007), a listwise ranking approach, which directly optimizes the evaluation metric, averaged over all mentions in the training data. The details about RankSVM model is discussed in Appendix A.3.

Recent neural network models also mostly use a ranking objective. For instance, the softmax loss is a common choice (Logeswaran et al. 2019; Wu et al. 2020; Botha et al. 2020). Conceptually, the score for the target entity is maximized while the sum of scores for other entities in the candidate set or some negative sampled set is minimized.

6.4.1.3 Structured Prediction

Binary classification and learning to rank models usually learn local scoring functions that are used in the local inference (Eq. 6.3) or the approximated global inference (Eq. 6.5). In order to learn a global scoring function, structured prediction models are needed. Structured prediction models make predictions for multiple mentions simultaneously. When training such models, the main challenge is again the scaling problem that we saw in global inference. Since the label space grows exponentially with the number of mentions, inference speed is usually much slower and more training instances may be needed for models to generalize well.

Some work assume simple dependencies among mentions, such as linear or tree structures. That is, the decision of a mention only depends on some previous mentions in the document. This assumption enjoys efficient inference algorithm based on dynamic programming. For example, Nguyen et al. (2016) jointly model NER and entity linking by concatenating NER labels and Wikipedia titles. For instance, the modified label for a token can be "PER:Barck_Obama", where "PER" is the named entity type and "Barack_Obama" is the corresponding Wikipedia title. Besides the linear-chain model, they also propose a tree model in which a factor is added between two tokens if they are connected according to a dependency parser. The exact inference is still tractable by variants of the Viterbi algorithm.

If no simple dependency is assumed, approximation inference techniques are required to make inference tractable. In the following, we briefly discuss some example works. In Nguyen et al. (2016), in addition to the assumption of linear and tree structures, they propose another global model in which mentions across sentences are linked if they share any candidate entity. Since the dependency structure would not be a tree now, exact inference is intractable. They propose to apply Gibbs sampling (Finkel et al. 2005) to approximate the solution. Guo et al. (2013) use structural SVM to jointly perform mention detection and disambiguation on tweets. Because there are second-order features between mentions, they order the mentions from left to right, and use beam search algorithm to find the joint assignment approximately. Ganea et al. (2016) propose a probabilistic graphical model which is essentially a complete graph since each mention depends on all other mentions. They use heuristics to prune the number of variables and apply loopy belief propagation to perform approximate inference. Yang et al. (2018) employ a structured gradient boosting tree algorithm for disambiguating entity mentions collectively. They propose an approximate inference algorithm named Bidirectional Beam Search with Gold path, which leverages global information from both past and future to perform better search.

6.4.1.4 Generative Models

Unlike the above feature-based models, models in this category assume a generative story for the mentions in the documents, and maximize the likelihood of observing mentions, the corresponding entities, and context words during training. Each model makes different assumptions of how the labels are generated, namely, the dependency between variables and how the joint probability is decomposed. We give two examples below.

Han and Sun (2011) propose an entity-mention generative model. For each document, the entities are first chosen according to some popularity knowledge which gives the likelihood of an entity appearing in a document. Based on these entities, the mention surface strings and the context around each mention are then generated. Han and Sun (2012) further incorporate the idea of topical coherence and propose a topic-mention model. This model modifies the topic models by adding entities and entity mentions into the document generation process. For each document, the model first draws a topic, and then entities are generated according to the topic. Finally, mentions and other words in the document are generated based on the drawn entities.

6.4 Learning Scoring Functions

Lazic et al. (2015) propose a selective context model which assumes that most features that appear in the context are not discriminative. They propose to use a latent variable to select a single contextual feature for each mention. The features in this work are other name mentions and noun phrases in the same document. The document generation process is as follow. Mentions are first chosen according to some prior probability. For each mention, the corresponding entity is then generated based on a prior distribution over possible entities for a mention. The latent variable selects a relevant context feature that fires for this entity. Finally, the remaining features are drawn from a background distribution.

Their proposed generative model can be applied to the semi-supervised setting naturally, in which the goal is to improve a supervised model using large amount of unlabeled data. In their model, there is a variable indicates the probability of an entity given a mention. If a mention is not labeled in the training data, the EM algorithm will calculate the expected likelihood using this variable. Otherwise, the probability mass is simply set to the ground truth entity. In their experiments, they use Wikipedia anchor texts as the labeled data, and use a Web corpus of 50 million pages as the unlabeled data. They achieve substantial improvement by adding unlabeled data when the model is evaluated on the CoNLL-AIDA dataset.

6.4.2 Unsupervised Methods

Unsupervised methods do not require labeled examples to train a model. Although the anchor texts in Wikipedia provide free training data for entity linking with Wikipedia, several researchers develop unsupervised approaches with aims to be robust on different domains or to enhance an supervised model. However, if there are enough in-domain training data, supervised methods usually can achieve a better performance. Since most research still use Wikipedia or Wikidata as the target KB, unsupervised approaches are less popular, especially in the reent deep learning era.

6.4.2.1 Similarity-Based Methods

A common approach in unsupervised methods is to build a representation for a candidate entity and a representation for the corresponding mention, and then choose the most similar candidate according to these representations. These representations do not require any training examples for the entity linking task.

Cucerzan (2007) employs a vector space model, in which the vector representation of the document is compared with the vector representation of candidate entity. The entity which has the highest similarity will be selected as the answer. Wikipedia categories and contexts in the Wikipedia page are used to represent an entity. A document is represented by all other mentions and context words.

Pan et al. (2015) construct a knowledge network for each mention based on Abstract Meaning Representation (AMR) (Banarescu et al. 2013) of the document. The key idea is that AMR can be used to select more important mentions in the context for the target mention, since AMR provides richer analysis on text such as entity typing, coreference, and some semantic roles. This mention knowledge network is then compared with each candidate's knowledge network which is based on the relations in Wikipedia, DBpedia, and Freebase. If a candidate's network has more overlap with the mention's knowledge network, it will be ranked higher.

The above two models essentially perform local inference. Global inference can also be achieved in a unsupervised model. Similar to Pan et al. (2015), Wang et al. (2015) construct a graph for all mentions in a documents, and disambiguate all mentions jointly. In this graph, two mentions are connected if they are in a proximity or if there is a coreference relation. For each combination of candidate assignment in the mention graph, a candidate graph is built based on the relations between the selected candidates in the KB. Finally, each candidate graph is compared with the mention graph in order to select the best candidate assignments.

6.4.2.2 Graph-Based Methods

Graph-based approaches construct a graph using mentions and candidate entities as nodes, and the edges between two nodes are based on the existence of relationships or compatibility of the two nodes. This type of models usually perform global inference since decisions of multiple mentions are made simultaneously.

Han et al. (2011) use n-grams which are more likely to be keyphrases as the mention nodes and initialize their scores based on the TF-IDF measure. A random walk with restart algorithm is proposed to propagate these initial scores to title candidate nodes. Moreover, these scores will be propagated proportional to the weights on the edges. The edge weights are computed from mention-entity similarity or entity-entity similarity features.

Moro et al. (2014) build a graph for each document using BabelNet,[6] a semantic network constructed from multilingual Wikipedia, WordNet, and other resources. Instead of directly using the relations in BabelNet, for each candidate entity, they perform a random walk on BabelNet, starting from this entity. The results will be used to gather a semantic signature for each entity, which basically contains strongly related entities. Given a document, the proposed graph consists of nodes from the candidates of mentions, and edges created according to the pre-computed semantic signatures. They further propose a densest subgraph heuristic to reduce the level of ambiguity and select the final answers.

Hoffart et al. (2011) propose a graph-based model on top of scores from a supervised model. They construct a mention-entity graph in which nodes are mentions and candidates. Each mention-entity edge is weighted by a combination of entity popularity features and context similarity features. In addition, Each entity-entity edge is weighted by the link-based similarity between two Wikipedia pages. Given this graph, they propose an algorithm to compute a dense subgraph that contains exactly one mention-entity edge for each mention.

---

[6] http://babelnet.org.

6.5 NIL Mention Identification

In the previous sections, we discussed approaches for choosing the best entity from the candidate set. However, the target entity does not always exist in the knowledge base since the KB is usually not perfect. Only popular or prominent entities will be added to a knowledge base. For such cases, if the candidate generation algorithm still creates a non-empty candidate set, the top-ranked entity is guaranteed to be a wrong answer. The goal of this step is therefore to identify this kind of mentions and output "NIL" as the answer for them.

Most studies of entity linking simply do not handle this issue and assume target entities are always in the given knowledge base. In this case, NIL mentions are the mentions which have zero candidate entities.

One simple approach is to employ a threshold on the inference scores (Bunescu and Paşca 2006). Namely, if the inference score of the top-ranked candidate is lower than a threshold, "NIL" will be returned instead. This threshold could be learnt from training data or manually tuned.

Another popular approach is to learn a supervised binary classifier to decide if the top-ranked entity should be rejected (Zheng et al. 2010). The features used in this binary classifier are vastly identical to the ones we introduced in Sect. 6.2. Besides only using information from the mention and the corresponding top candidate, Ratinov et al. (2011) design additional features based on the top candidate and the runner-up.

Some systems do not have an additional step for NIL mention identification. Instead, "NIL" is added to the candidate set of each mention as a special entity, so the NIL mention detection problem is jointly solved with context sensitive inference. For example, Dredze et al. (2010) add an additional "NIL" candidate for each mention, and design features which capture properties which indicate if the mention could be NIL. For example, if there is no candidate has a good name similarity with the mention, this mention is likely to be NIL. whether any of the candidates match the mention. The features for this NIL candidate could be statistics taken on feature values of other candidate entities. Guo et al. (2013) also add NIL as one of the candidate for each mention candidate. In their structural SVM model, a special bias feature is assigned to each mention, and the learned weight of this bias term will be used as the threshold to cut-off mentions. Since their model jointly perform mention detection and linking, if a candidate mention is linked to NIL, it will not be treated as a mention in their final outputs. In the generative model proposed by Han and Sun (2011), they use the idea that if a mention refers to an specific entity, the probability of this mention generated by this entity's model should be much higher than the probability of it is generated by a general language model. They add a NIL entity into the knowledge base and assume that the NIL entity generates mentions according to the general language model.

6.6 Discussion and Conclusions

In this chapter, we looked at the problem of choosing the most suitable entity for each mention in a document. The linking problem can be formulated as an optimization problem, in which entities are chosen so that an objective function which measures the compatibility between mentions and entities is maximized.

We started by discussing features that could represent mentions in the context and represent entities in a knowledge base. These features are the building blocks of the objective function. Regardless of traditional machine learning models or deep learning models, good representations of mentions and entities are the key to an accurate model. Most models only use local features, in which each pair of mention and entity is measured independently of other mentions and entities. In contrast, the more complex global features consider multiple mentions and their entities simultaneously, so they can leverage relationships between entities. Although global features are more expressive and powerful in theory, most entities can be disambiguated correctly only using local features. Therefore, the effectiveness of global features might not be shown in most datasets. For the harder cases, global information is still needed.

Feature generation is the place where the cross-lingual challenge comes into play. Because mentions and entities are written in two different languages in the cross-lingual entity linking setup, features need to be able to capture the compatibility and similarity across languages. When Wikipedia is the target knowledge base, the structure of inter-language links again becomes a valuable source for addressing this issue. More recently, pre-trained multilingual language models have been shown to be very useful for solving the cross-lingual challenges.

Given functions that measures compatibility between mentions and entities, either local or global inference models can be used to select entities which maximized the overall score. A local inference model makes decision on each mention independently, whereas a global inference model leverages global features to resolve multiple mentions simultaneously. The global inference problem is NP-hard in general, since the number of entity combinations grows exponentially with the number of mentions. We discussed several approximation inference strategies that decompose the inference problem into smaller but tractable sub-problems.

Finally, we looked at how these scoring functions are learnt. Supervised learning methods is predominantly used in the literature, since Wikipedia and Wikidata are the most popular target knowledge bases and there are tons of free annotations on the web for them. Among supervised approaches, most models use some ranking loss regardless of neural or non-neural network models. A model will assign a score to each entity candidate, and the model parameters are learnt so that the score of the correct entity is maximized while the scores of other candidates are minimized.

6.7 Bibliographical Notes

Local Knowledge Base Features Almost all entity linking systems (Hoffart et al. 2011; Kulkarni et al. 2009; Ratinov et al. 2011; Shen et al. 2012; Ling et al. 2015; Luo et al. 2015; Nguyen et al. 2016; Yamada et al. 2016; Tsai and Roth 2016b) use some entity popularity features. Besides the prior probability of Eq. 6.2, (Gattani et al. 2013) also use view statistics of Wikipedia pages as features. Dredze et al. (2010) look at the number of Wikipedia pages links to and from a candidate entity to represent the popularity of the candidate. For name similarity features, (Zheng et al. 2010; Liu et al. 2013) use edit distance to compare mentions and entity names, whereas (Dredze et al. 2010) employ character Dice and Hamming distance.

Local Contextual Similarity Features The context of a mention could be the words in the entire document (Guo et al. 2013; Liu et al. 2013), words in a small window around the mention (Bunescu and Paşca 2006; Kulkarni et al. 2009; Han et al. 2011; Ratinov et al. 2011), or other mentions in the document (Hoffart et al. 2011; Zhang et al. 2010; Dredze et al. 2010). For entries in Wikipedia, the context could be the words in the entire Wikipedia page (Bunescu and Paşca 2006; Kulkarni et al. 2009; Ratinov et al. 2011; Han et al. 2011), or words in the first paragraph (Kulkarni et al. 2009).

Global Features Cucerzan (2007) uses the overlap between Wikipedia categories and words in the Wikipedia pages to measure the coherency between two Wikipedia pages. Milne and Witten (2008) propose the link-based measure. The idea is that two Wikipedia pages are more semantically related if there are more Wikipedia articles that link to both of them. Besides the incoming links to the titles, Ratinov et al. (2011) also use outgoing links to measure similarity between two titles. Given the sets of incoming links or outgoing links of two pages, Milne and Witten (2008) use Normalized Google Distance for the similarity measure. Ratinov et al. (2011) further calculate PMI (Point-wise Mutual Information) between the two sets of links, whereas Guo et al. (2013) use Jaccard similarity. For the less popular entities, there might not be enough incoming links, outgoing links, and categories for measuring robust similarity. Hoffart et al. (2012) propose to extract keyphrases in the Wikipedia article, and compute similarity between two pages based on the overlap of keyphrases. Globerson et al. (2016) measure similarity between two Wikipedia pages using the number of Freebase relations between them, the number of hyperlinks between the two Wikipedia pages (in either direction), and the number of common mentions in the two pages.

Inference Models Local inference is a popular approach (Mihalcea and Csomai 2007; Durrett and Klein 2014; Lazic et al. 2015; Francis-Landau et al. 2016), owing to its simplicity.

For the work that use approximated global inference, Cucerzan (2007) uses candidates of all other mentions in the document to represent $G(m_i)$ in Eq. 6.5. Tsai and Roth (2016b) only take the entities from the previously disambiguated mentions to generate entity coherence features. Similar to Ratinov et al. (2011)'s two-stage approach, Globerson et al. (2016) take the most "supportive" candidate from each mention in the document as $G(m_i)$. The support is measured by similarity between entities and a proposed attention model.

Supervised Methods Dredze et al. (2010), Yamada et al. (2016), Chen and Ji (2011), Kulkarni et al. (2009), Shen et al. (2012), Zhang et al. (2011), Chisholm and Hachey (2015) also use RankSVM model to learn local scoring functions.

Unsupervised Methods Besides the pure unsupervised methods we discussed in Sect. 6.4.2, some approaches add an additional unsupervised inference upon the results of a supervised model. In other words, the scores obtained by a supervised model are used in an unsupervised method. Kulkarni et al. (2009) try to select the best candidate for each mention jointly via maximizing an objective which consists of a local compatibility score and a label relatedness score. The local compatibility score for each candidate is obtained from a rankSVM model, and the label relatedness score comes from some similarity measures between two entities. Since exact inference over all candidate assignments is intractable, they propose an approach based on local hill-climbing and rounding integer linear programs. Similarly, in order to choose the best entity assignments for all mentions simultaneously, Cheng and Roth (2013) formulate a Constrained Conditional Model (Roth and Yih 2004; Chang et al. 2012) using Integer Linear Programming. The model consists of two parts. The first part contains ranking scores from Ratinov et al. (2011)'s supervised ranker, and the second part contains relational scores for each pair of candidates from two mentions. Similar to the graph-based approach proposed in Han et al. (2011), Pershina et al. (2015) perform a variant of personalized PageRank algorithm on the graph in which nodes are candidate entities in a document, and an edge exists if the two entities are linked in Wikipedia. Besides the graph construction is different from Han et al. (2011), they initialize the score of each node by a supervised classifier.

NIL Mention Identification Several studies (Cucerzan 2007; Kulkarni et al. 2009; Han et al. 2011; Pershina et al. 2015; Yamada et al. 2016) simply do not handle this issue and assume target entities are in the given knowledge base. Besides Pilz and Paaß (2011), Bunescu and Paşca (2006), Han and Sun (2012), Shen et al. (2012), Gottipati and Jiang (2011), Ferragina and Scaiella (2010), Li et al. (2013) also employ a threshold on the inference scores. If the top-ranked entity is lower than this threshold, return NIL instead. Zhang et al. (2011), Han and Sun (2011) also learn a supervised binary classifier to decide if the top-ranked entity should be rejected. Luo et al. (2015), Nguyen et al. (2016) also incorporate the NIL mention detection problem into ranking models

References

Dredze, M., McNamee, P., Rao, D., Gerber, A., Finin, T.: Entity disambiguation for knowledge base population. In: Proceedings of the 23rd International Conference on Computational Linguistics (Coling 2010), Coling 2010 Organizing Committee, pp. 277–285. Beijing, China (2010). https://aclanthology.org/C10-1032

Guo, S., Chang, M.-W., Kicimanm, E.: To link or not to link? a study on end-to-end tweet entity linking. In: Proceedings of the 2013 Conference of the North American Chapter of the Association for Computational Linguistics: Human Language Technologies, Association for Computational Linguistics, pp. 1020–1030. Atlanta, Georgia (2013). https://aclanthology.org/N13-1122

Ling, X., Singh, S., Weld, D.S.: Design challenges for entity linking. Trans. Assoc. Comput. Linguistics **3**, 315–328 (2015)

Ratinov, L., Roth, D., Downey, D., Anderson, M.: Local and global algorithms for disambiguation to Wikipedia. In: Proceedings of the 49th Annual Meeting of the Association for Computational Linguistics: Human Language Technologies, Association for Computational Linguistics, pp. 1375–1384. Portland, Oregon, USA (2011). https://aclanthology.org/P11-1138

Yamada, I., Shindo, H., Takeda, H., Takefuji, Y.: Joint learning of the embedding of words and entities for named entity disambiguation. In: Proceedings of the 20th SIGNLL Conference on Computational Natural Language Learning, Association for Computational Linguistics, pp. 250–259. Berlin, Germany (2016). https://doi.org/10.18653/v1/K16-1025, https://aclanthology.org/K16-1025

Francis-Landau, M., Durrett, G., Klein, D.: Capturing semantic similarity for entity linking with convolutional neural networks. In: Proceedings of the 2016 Conference of the North American Chapter of the Association for Computational Linguistics: Human Language Technologies, Association for Computational Linguistics, pp. 1256–1261. San Diego, California (2016). https://doi.org/10.18653/v1/N16-1150, https://aclanthology.org/N16-1150

He, Z., Liu, S., Li, M., Zhou, M., Zhang, L., Wang, H.: Learning entity representation for entity disambiguation. In: Proceedings of the 51st Annual Meeting of the Association for Computational Linguistics (Volume 2: Short Papers), Association for Computational Linguistics, pp. 30–34. Sofia, Bulgaria (2013). https://aclanthology.org/P13-2006

Gupta, N., Singh, S., Roth, D.: Entity linking via joint encoding of types, descriptions, and context. In: Proceedings of the 2017 Conference on Empirical Methods in Natural Language Processing, Association for Computational Linguistics, pp. 2681–2690. Copenhagen, Denmark (2017). https://doi.org/10.18653/v1/D17-1284, https://aclanthology.org/D17-1284

Logeswaran, L., Chang, M.-W., Lee, K., Toutanova, K., Devlin, J., Lee, H.: Zero-shot entity linking by reading entity descriptions. In: Proceedings of the 57th Annual Meeting of the Association for Computational Linguistics, Association for Computational Linguistics, pp. 3449–3460. Florence, Italy (2019). https://doi.org/10.18653/v1/P19-1335, https://aclanthology.org/P19-1335

Wu, L., Petroni, F., Josifoski, M., Riedel, S., Zettlemoyer, L.: Scalable zero-shot entity linking with dense entity retrieval. In: Proceedings of the 2020 Conference on Empirical Methods in Natural Language Processing (EMNLP), Association for Computational Linguistics, pp. 6397–6407. Online (2020). https://doi.org/10.18653/v1/2020.emnlp-main.519, https://aclanthology.org/2020.emnlp-main.519

Li, B.Z., Min, S., Iyer, S., Mehdad, Y., Yih, W.-T.: Efficient one-pass end-to-end entity linking for questions. In: Proceedings of the 2020 Conference on Empirical Methods in Natural Language Processing (EMNLP), Association for Computational Linguistics, pp. 6433–6441. Online (2020a). https://doi.org/10.18653/v1/2020.emnlp-main.522, https://aclanthology.org/2020.emnlp-main.522

Tsai, C.-T., Roth, D.: Cross-lingual wikification using multilingual embeddings. In: Proceedings of the 2016 Conference of the North American Chapter of the Association for Computational

Linguistics: Human Language Technologies, Association for Computational Linguistics, pp. 589–598. San Diego, California (2016b). https://doi.org/10.18653/v1/N16-1072, https://aclanthology.org/N16-1072

Upadhyay, S., Gupta, N., Roth, D.: Joint multilingual supervision for cross-lingual entity linking. In: Proceedings of the 2018 Conference on Empirical Methods in Natural Language Processing, Association for Computational Linguistics, pp. 2486–2495. Brussels, Belgium (2018). https://doi.org/10.18653/v1/D18-1270, https://aclanthology.org/D18-1270

Smith, S.L., Turban, D.H., Hamblin, S., Hammerla, N.Y.: Offline bilingual word vectors, orthogonal transformations and the inverted softmax. In: Proceedings of the International Conference on Learning Representations (ICLR) (2017)

Botha, J.A., Shan, Z., Gillick, D.: Entity Linking in 100 Languages. In: Proceedings of the 2020 Conference on Empirical Methods in Natural Language Processing (EMNLP), Association for Computational Linguistics, pp. 7833–7845. Online (2020). https://doi.org/10.18653/v1/2020.emnlp-main.630, https://aclanthology.org/2020.emnlp-main.630

Cheng, X., Roth, D.: Relational inference for wikification. In: Proceedings of the 2013 Conference on Empirical Methods in Natural Language Processing, Association for Computational Linguistics, pp. 1787–1796. Seattle, Washington, USA (2013). https://aclanthology.org/D13-1184

Cucerzan, S.: Large-scale named entity disambiguation based on Wikipedia data. In: Proceedings of the 2007 Joint Conference on Empirical Methods in Natural Language Processing and Computational Natural Language Learning (EMNLP-CoNLL), Association for Computational Linguistics, pp. 708–716. Prague, Czech Republic (2007). https://aclanthology.org/D07-1074

Ganea, O.-E., Ganea, M., Lucchi, A., Eickhoff, C., Hofmann, T.: Probabilistic bag-of-hyperlinks model for entity linking. In: Proceedings of the 25th International Conference on World Wide Web (WWW), pp. 927–938 (2016)

Globerson, A., Lazic, N., Chakrabarti, N., Subramanya, A., Ringgaard, M., Pereira, F.: Collective entity resolution with multi-focal attention. In: Proceedings of the 54th Annual Meeting of the Association for Computational Linguistics (Volume 1: Long Papers), Association for Computational Linguistics, pp. 621–631. Berlin, Germany (2016). https://doi.org/10.18653/v1/P16-1059, https://aclanthology.org/P16-1059

Ganea, O.-E., Hofmann, T.: Deep joint entity disambiguation with local neural attention. In: Proceedings of the 2017 Conference on Empirical Methods in Natural Language Processing, Association for Computational Linguistics, pp. 2619–2629. Copenhagen, Denmark (2017). https://doi.org/10.18653/v1/D17-1277, https://aclanthology.org/D17-1277

Yang, Y., Chang, M.-W.: S-MART: novel tree-based structured learning algorithms applied to tweet entity linking. In: Proceedings of the 53rd Annual Meeting of the Association for Computational Linguistics and the 7th International Joint Conference on Natural Language Processing (Volume 1: Long Papers), Association for Computational Linguistics, pp. 504–513. Beijing, China (2015). https://doi.org/10.3115/v1/P15-1049, https://aclanthology.org/P15-1049

Milne, D., Witten, I.H.: Learning to link with Wikipedia. In: Proceedings of the ACM Conference on Information and Knowledge Management (CIKM) (2008)

Pilz, A., Paaß, G.: From names to entities using thematic context distance. In: Proceedings of the ACM Conference on Information and Knowledge Management (CIKM), pp. 857–866 (2011)

Zhang, W., Su, J., Tan, C.L., Wang, W.T.: Entity linking leveraging automatically generated annotation. In: Proceedings of the 23rd International Conference on Computational Linguistics (Coling 2010), Coling 2010 Organizing Committee, pp. 1290–1298. Beijing, China (2010). https://aclanthology.org/C10-1145

Sil, A., Yates, A.: Re-ranking for joint named-entity recognition and linking. In: Proceedings of the 22nd ACM International Conference on Information & Knowledge Management (CIKM), pp. 2369–2374. ACM (2013)

Zheng, Z., Li, F., Huang, M., Zhu, X.: Learning to link entities with knowledge base. In: Human Language Technologies: The 2010 Annual Conference of the North American Chapter of the Association for Computational Linguistics, Association for Computational Linguistics, pp. 483–491. Los Angeles, California (2010). https://aclanthology.org/N10-1072

Bunescu, R., Paşca, M.: Using encyclopedic knowledge for named entity disambiguation. In: 11th Conference of the European Chapter of the Association for Computational Linguistics, pp. 9–16. Trento, Italy (2006). https://aclanthology.org/E06-1002

Chen, Z., Ji, H.: Collaborative ranking: A case study on entity linking. In: Proceedings of the 2011 Conference on Empirical Methods in Natural Language Processing, Association for Computational Linguistics, pp. 771–781. Edinburgh, Scotland, UK (2011). https://aclanthology.org/D11-1071

Cao, Z., Qin, T., Liu, T., Tsai, M., Li, H.: Learning to rank: from pairwise approach to listwise approach. In: Ghahramani, Z. (ed.) Proceedings of the International Conference on Machine Learning (ICML) (2007)

Nguyen, D.B., Theobald, M., Weikum, G.: J-NERD: joint named entity recognition and disambiguation with rich linguistic features. Trans. Assoc. Comput. Linguistics **4**, 215–229 (2016)

Finkel, J.R., Grenager, T., Manning, C.: Incorporating non-local information into information extraction systems by Gibbs sampling. In: Proceedings of the 43rd Annual Meeting of the Association for Computational Linguistics (ACL'05), Association for Computational Linguistics, pp. 363–370. Ann Arbor, Michigan (2005). https://doi.org/10.3115/1219840.1219885, https://aclanthology.org/P05-1045

Yang, Y., Irsoy, O., Rahman, K.S.: Collective entity disambiguation with structured gradient tree boosting. In: Proceedings of the 2018 Conference of the North American Chapter of the Association for Computational Linguistics: Human Language Technologies, Volume 1 (Long Papers), Association for Computational Linguistics, pp. 777–786. New Orleans, Louisiana (2018). https://doi.org/10.18653/v1/N18-1071, https://aclanthology.org/N18-1071

Han, X., Sun, L.: A generative entity-mention model for linking entities with knowledge base. In: Proceedings of the 49th Annual Meeting of the Association for Computational Linguistics: Human Language Technologies, Association for Computational Linguistics, pp. 945–954. Portland, Oregon, USA (2011). https://aclanthology.org/P11-1095

Han, X., Sun, L.: An entity-topic model for entity linking. In: Proceedings of the 2012 Joint Conference on Empirical Methods in Natural Language Processing and Computational Natural Language Learning, Association for Computational Linguistics, pp. 105–115. Jeju Island, Korea (2012). https://aclanthology.org/D12-1010

Lazic, N., Subramanya, A., Ringgaard, M., Pereira, F.: Plato: a selective context model for entity resolution. Trans. Assoc. Comput. Linguistics **3**, 503–515 (2015)

Pan, X., Cassidy, T., Hermjakob, U., Ji, H., Knight, K.: Unsupervised entity linking with Abstract Meaning Representation. In: Proceedings of the 2015 Conference of the North American Chapter of the Association for Computational Linguistics: Human Language Technologies, Association for Computational Linguistics, pp. 1130–1139. Denver, Colorado (2015). https://doi.org/10.3115/v1/N15-1119, https://aclanthology.org/N15-1119

Banarescu, L., Bonial, C., Cai, S., Georgescu, M., Griffitt, K., Hermjakob, U., Knight, K., Koehn, P., Palmer, M., Schneider, N.: Abstract Meaning Representation for sembanking. In: Proceedings of the 7th Linguistic Annotation Workshop and Interoperability with Discourse, Association for Computational Linguistics, pp. 178–186. Sofia, Bulgaria (2013). https://aclanthology.org/W13-2322

Wang, H., Zheng, J.G., Ma, X., Fox, P., Ji, H.: Language and domain independent entity linking with quantified collective validation. In: Proceedings of the 2015 Conference on Empirical Methods in Natural Language Processing, Association for Computational Linguistics, pp. 695–704. Lisbon, Portugal (2015). https://doi.org/10.18653/v1/D15-1081, https://aclanthology.org/D15-1081

Han, X., Sun, L., Zhao, J.: Collective entity linking in web text: a graph-based method. In: Proceedings of the ACM SIGIR Conference (SIGIR), pp. 765–774 (2011)

Moro, A., Raganato, A., Navigli, R.: Entity linking meets word sense disambiguation: a unified approach. Trans. Assoc. Comput. Linguistics **2**, 231–244 (2014)

Hoffart, J., Yosef, M.A., Bordino, I., Fürstenau, H., Pinkal, M., Spaniol, M., Taneva, B., Thater, S., Weikum, G.: Robust disambiguation of named entities in text. In: Proceedings of the 2011 Conference on Empirical Methods in Natural Language Processing, Association for Computational Linguistics, pp. 782–792. Edinburgh, Scotland, UK (2011). https://aclanthology.org/D11-1072

Kulkarni, S., Singh, A., Ramakrishnan, G., Chakrabarti, S.: Collective annotation of Wikipedia entities in web text. In: Proceedings of the 15th ACM SIGKDD Conference on Knowledge Discovery and Data Mining (KDD), pp. 457–466. ACM (2009)

Shen, W., Wang, J., Luo, P., Wang, M.: LINDEN: linking named entities with knowledge base via semantic knowledge. In: Proceedings of the 21st International Conference on World Wide Web (WWW), pp. 449–458. ACM (2012)

Luo, G., Huang, X., Lin, C.-Y., Nie, Z.: Joint entity recognition and disambiguation. In: Proceedings of the 2015 Conference on Empirical Methods in Natural Language Processing, Association for Computational Linguistics, pp. 879–888. Lisbon, Portugal (2015). https://doi.org/10.18653/v1/D15-1104, https://aclanthology.org/D15-1104

Gattani, A., Lamba, D.S., Garera, N., Tiwari, M., Chai, X., Das, S., Subramaniam, S., Rajaraman, A., Harinarayan, V., Doan, A.: Entity extraction, linking, classification, and tagging for social media: a Wikipedia-based approach. Very Large Data Base (VLDB) Endowment **6**(11), 1126–1137 (2013)

Liu, X., Li, Y., Wu, H., Zhou, M., Wei, F., Lu, Y.: Entity linking for tweets. In: Proceedings of the 51st Annual Meeting of the Association for Computational Linguistics (Volume 1: Long Papers), Association for Computational Linguistics, pp. 1304–1311. Sofia, Bulgaria (2013). https://aclanthology.org/P13-1128

Hoffart, J., Seufert, S., Nguyen, D.B., Theobald, M., Weikum, G.: KORE: keyphrase overlap relatedness for entity disambiguation. In: Proceedings of the ACM Conference on Information and Knowledge Management (CIKM), pp. 545–554. ACM (2012)

Mihalcea, R., Csomai, A.: Wikify!: linking documents to encyclopedic knowledge. In: Proceedings of the ACM Conference on Information and Knowledge Management (CIKM) (2007)

Durrett, G., Klein, D.: A joint model for entity analysis: Coreference, typing, and linking. Trans. Assoc. Comput. Linguistics **2**, 477–490 (2014)

Zhang, W., Sim, Y.-C., Su, J., Tan, C.-L.: Entity linking with effective acronym expansion, instance selection and topic modeling. In: Twenty-Second International Joint Conference on Artificial Intelligence (IJCAI) (2011)

Chisholm, A., Hachey, B.: Entity disambiguation with web links. Trans. Assoc. Comput. Linguistics **3**, 145–156 (2015)

Roth, D., Yih, W.-T.: A linear programming formulation for global inference in natural language tasks. In: Proceedings of the Eighth Conference on Computational Natural Language Learning (CoNLL-2004) at HLT-NAACL 2004, Association for Computational Linguistics, pp. 1–8. Boston, Massachusetts, USA (2004). https://aclanthology.org/W04-2401

Chang, M.-W., Ratinov, L., Roth, D.: Structured learning with constrained conditional models. Machine Learn. **88**(3), 399–431 (2012). http://cogcomp.org/papers/ChangRaRo12.pdf

Pershina, M., He, Y., Grishman, R.: Personalized page rank for named entity disambiguation. In: Proceedings of the 2015 Conference of the North American Chapter of the Association for Computational Linguistics: Human Language Technologies, Association for Computational Linguistics, pp. 238–243. Denver, Colorado (2015). https://doi.org/10.3115/v1/N15-1026, https://aclanthology.org/N15-1026

Gottipati, S., Jiang, J.: Linking entities to a knowledge base with query expansion. In: Proceedings of the 2011 Conference on Empirical Methods in Natural Language Processing, Association for Computational Linguistics, pp. 804–813. Edinburgh, Scotland, UK (2011). https://aclanthology.org/D11-1074

Ferragina, P., Scaiella, U.: TAGME: on-the-fly annotation of short text fragments (by Wikipedia entities). In: Proceedings of the ACM Conference on Information and Knowledge Management (CIKM) (2010)

Li, Y., Wang, C., Han, F., Han, J., Roth, D., Yan, X.: Mining evidences for named entity disambiguation. In: Proceedings of the ACM SIGKDD Conference on Knowledge Discovery and Data Mining (KDD), pp. 1070–1078. ACM (2013)

Recent Advances and Future Directions 7

7.1 Recent Advances

Since around 2015, many approaches for entity linking started to leverage neural network or deep learning models, pushing the state-of-the-art performance on this task. We have introduced some of them in the previous chapters. In this section, we highlight two recent threads of work: dense entity retrieval and auto-regressive entity retrieval.

7.1.1 Dense Entity Retrieval

Entity linking can be viewed as an information retrieval problem. The query is a mention in text, and the goal is to retrieve a Wikipedia article that the mention refers to. Traditional retrieval methods such as TF-IDF and BM25 are built on sparse representations of both queries and documents. Recently, several works have shown that using dense representations could achieve a superior performance on a range of retrieval tasks (Reimers and Gurevych 2019; Gillick et al. 2019; Hofstätter et al. 2021). This dense retrieval idea has been successfully applied to entity linking, as we have seen in retrieval-based candidate generation method in Sect. 5.1.3. In this section, we will discuss Wu et al. (2020) in more details.

7.1.1.1 Bi-Encoder
Wu et al. (2020) propose a bi-encoder architecture for generating entity candidates. Let x be a textual input (a sentence with an annotated mention) and e be an entity in the knowledge base, they are encoded into dense vectors:

$$\mathbf{y_x} = reduce(T_1(x))$$
$$\mathbf{y_e} = reduce(T_2(e)),$$

where T_1 and T_2 are two independent transformer models and $reduce$ is a function that reduces a sequence of vectors into one vector. The score of this entity candidate is simply the dot-product between the two dense representations:

$$score(x, e) = \mathbf{y_x} \cdot \mathbf{y_x}.$$

The model is trained to maximize the scores of the correct entities with respect to randomly sampled negative entities.

The input x to the mention encoder is "[CLS] ctxt$_l$ [M$_s$] mention [M$_e$] ctxt$_r$ [SEP]", where ctxt$_l$ and ctxt$_r$ are the word-pieces tokens of the left and right context, and [M$_s$] and [M$_e$] are special tokens to tag the start and end of the mention. The input to the entity encoder e is "[CLS] title [ENT] description [SEP]", where title and description are word-pieces tokens of the entity title and this entity's description. For both mention and entity encoders, the $reduce$ function simply takes the "[CLS]" token from the last layer of transformer models.

Since they do not use a dictionary-based approach to pre-generate a small set of entity candidates, the inference problem becomes finding the maximum dot product between the mention representation and all entity representations in the knowledge base. One advantage of this architecture is that entity representations can be pre-computed and cached. In addition, in order to improve the efficiency of retrieving top k similar dense vectors, they utilize the exact and approximate nearest neighbor search algorithms implemented in FAISS (Johnson et al. 2019). In their experiments, an exact search method takes 9.2 ms on average to return top 100 candidates per query. An approximate search reduces the average query time to 2.6 ms with less than 1.2% drop in accuracy and recall.

In the experiments, the authors show significant recall improvement over dictionary-based and sparse retrieval approaches on both zero-shot entity linking and TACKBP-2010 datasets.

7.1.1.2 Cross-Encoder

After using the bi-encoder model to retrieve top k ($k \leq 100$) candidate entities, the authors use a more expressive cross-encoder model to re-rank the candidates. The input to the model is a concatenation of x and e, that is, "[CLS] ctxt$_l$ [M$_s$] mention [M$_e$] ctxt$_r$ [SEP] title [ENT] description [SEP]". This input is encoded by a transformer model T_{cross} and reduced to a single vector:

$$\mathbf{y_{x,e}} = reduce(T_{cross}(concat(x, e))).$$

Since the input sentence and entity description are concatenated, the transformer model can have deep cross attention between the two. Finally, a linear layer **W** is applied to obtain the score between x and e:

$$score_{cross}(x, e) = \mathbf{y_{x,e}W}.$$

7.1.1.3 Multilingual Variant

Botha et al. (2020) adapt the bi-encoder model to multilingual entity linking problem. The key difference here is that the transformer encoder is initialized from a pre-trained multilingual BERT checkpoint. Since in the multilingual setup, the textual input x and the entity description e are in two different languages, multilingual embeddings are required to generate a meaningful score between the two. For the inputs to the mention encoder, besides mention surface string, left context, and right context, document title is also included. Another difference from Wu et al. (2020) is that cosine similarity is used as the metric for scoring a mention-entity pair. This model is trained on anchor texts from 104 languages in Wikipedia.

In the experiments, the authors show significant improvement over existing models on several cross-lingual entity linking datasets, even the proposed model considers a much larger set of entity candidates and can handle more languages. The authors also provide preliminary results on using a cross-encoder to re-rank top k candidates retrieved by the bi-encoder model. Although there are some positive findings, this direction requires further research.

7.1.2 Autoregressive Entity Retrieval

As we have seen so far, most approaches formulate entity linking as a multi-class classification or ranking problem, in which each entity has its own representation (no matter sparse or dense). Entity representations are compared with mention representation in order to get some relevance score.

Instead of using this common viewpoint, Cao et al. (2021) solve entity linking as an autoregressive problem. Formally, let x be a textual input (a sentence with an annotated mention), an entity e in the knowledge base $\mathcal{K}$ is scored by an autoregressive formulation:

$$score(e|x) = p_\theta(y|x) = \prod_{i=1}^{N} p_\theta(y_i|y_{<i}, x),$$

where y is a set of N tokens that uniquely represents e and θ is model parameters. In other words, given the textual input x (a sentence with a mention), they use a sequence-to-sequence model to generate textual representations of entities. For instance, for an input $x = $ "Superman saved [START] Metropolis [END]", the desired output y is "Metropolis (comics)", assuming the target knowledge base is Wikipedia.

The proposed model optimizes a typical objective used for neural machine translation. More specifically, it maximizes the log likelihood log $p_\theta(y|x)$ with respect to model's parameters θ. At inference time, instead of computing a score for every $e \in \mathcal{K}$, the authors use Beam Search technique to search for the top-k entities with k beams. This approach is efficient since the length of entity representation y (Wikipedia title) is tractable and the k used in their experiment is small. Furthermore, allowing the model to generate any token from the vocabulary might lead to an output y that does not match any entity identifier in the knowledge base. The authors constrain the Beam Search process to only generate valid entity identifiers using a prefix tree constructed from all titles in Wikipedia.

The authors further extend this autoregression formulation to address end-to-end entity linking in which mention detection and entity disambiguation are performed jointly. For instance, given a sentence without any annotation, "In 1503, Leonardo began painting the Mona Lisa", the proposed sequence-to-sequence model would output "In 1503, [Leonardo](Leonardo da Vinci) began painting the Mona Lisa". The square brackets "[]" indicate the mention boundaries and the strings inside the parentheses "()" are entity identifiers. We can see that the model not only identifies the boundaries of mentions but also generates the corresponding Wikipedia titles for these mentions.

In their experiments, the proposed model performs comparably with the previous best model on entity disambiguation tasks. For end-to-end entity linking, the proposed model can outperform the state-of-the-art models significantly on some datasets. A key strength of this autoregressive formulation is that it has a significant reduction of memory footprint. Comparing to a dense retrieval method, the autoregressive method uses an order of magnitude less parameters since it only needs to store the prefix tree of entity names as opposed to a dense vector for every entity.

7.1.2.1 Multilingual Variant

De Cao et al. (2022) apply autoregressive formulation on the multilingual entity linking problem, in which the target knowledge base is multilingual (e.g., Wikidata). The assumption is that an entity has a unique textual identifier in at least one language. Therefore, the key research question addressed in the paper is how to generate entities' textual identifier given that there are multiple options in different languages.

The authors compared three strategies. The simplest one is to choose a single textual identifier among all the available languages of an entity. A downside of this approach is the chosen language might be quite different from the language of the input document. The model may not be able to exploit the lexical overlap between the context and the entity name efficiently. To address this issue, the second idea is to predict entity names in any language. That is, the model will generate a language identifier l and the entity name in language l:

$$score(e|x) = p_\theta(l|x) p_\theta(n_e^l|x, l), \qquad (7.1)$$

where n_e^l represents the textual identifier of entity e in language l. At inference time, the model can generate in any available language and therefore exploit synergies between the source and target language. The third idea treats the textual identifiers as a latent variable and marginalize over names in all languages:

$$score(e|x) = p_\theta(e|x) = \sum_{(l,n_e^l)} p_\theta(l, n_e^l|x).$$

However, marginalizing over all languages in training could be too expensive to compute. In this case, one could train the model with previous strategies and only apply the marginalization idea during inference time. In their experiments, the authors show that the model trains with Eq. (7.1) and with the marginalization idea in inference outperforms a dense retrieval method on several benchmarks.

One interesting finding in the paper is that a candidate generation step helps in the multilingual setup. That is, the authors apply a lookup-based candidate generation approach to reduce the search space greatly. Since in the multilingual setup, there are much more possible candidates for a mention, instead of ranking all of them, a simple candidate generation method could constrain the Beam Search steps and make the entire process more feasible.

One of the future directions for multilingual autoregression entity linking is to perform entity discovery jointly, such as in the monolingual case. Since mention detection in any language is already a very challenging problem, it would be interesting to see if entity disambiguation could help under the autoregression formulation.

7.2 Future Directions

Earlier chapters have discussed the development of Entity Linking and Multilingual Entity Linking over the last 15 years or so. The key trend observed is that the formulation of the problem has been simplified over the years and has become akin to an entity retrieval problem – representations of millions of entities can be stored and, given a mention in context, retrieved. The key lesson we can draw is that the basic task can be solved when mentions are given and there is enough supervision to learn good representations for the corresponding entities. Future work will have to go beyond this basic setting and address scenarios where data is scarce, target KBs vary (in content and languages) and the target mentions are more diverse, thus requiring more advanced inference method. We discuss below in more details some of these issues.

Low Resource Languages Recent efforts (Fu et al. 2020) have shown that the key step in the standard EDL pipeline, that of generating candidates from text, becomes very challenging when dealing with low resource languages. It is difficult especially when the entities are not represented in the native language KB. This is often the case since low resource languages also have smaller KBs. More work, and possibly the used of more resources ((Fu et al. 2020) suggested the use of query logs) will be needed to address these low-supervision cases.

Beyond Named Entities A second key area for future research has to do with moving beyond Named Entities. The are multiple kinds of non-named-entities that one should deal with. The simple case is that of linking mentions that are co-referred with named entities, but more involved cases, that require inference, exist. For example, ideally, a mention like the "US Allies" could like to various military alliances, possibly determined by the contextually relevant time period. This would require better context understanding and inference abilities that existing approaches are capable of. In particular, it would require better using the temporal context that has to be accounted for to link correctly. A further extension of this would move beyond name-entities and attempt to link events, such as WWII, or The Fall of the Berlin Wall, to their corresponding pages in Wikipedia. Another situation, which has been dealt with to some extent in the monolingual context, but not in the cross-lingual one, is the problem of mapping to NIL. This exists when a mention m in text has an identical (or highly similar) mention m that has a corresponding Wikipedia entity e, but m does not correspond to e or to any Wikipedia page. For example, in "My son's best friend in his 5th grade class is named John Kennedy." we do not want to link "John Kennedy" to any of the "John Kennedy" pages in Wikipedia, but rather to NIL. The last few problems mentioned bring up the issues of better use of context and, more generally, of global inference. While, as discussed earlier, a local approach can map a large fractions of the mentions correctly, global inference and better use of context are likely to be needed in the more subtle cases, when there is more variability in the text, and when we are expected to deal with NIL.

Additional KBs Essentially all the work in cross-lingual EDL has dealt with linking mentions to English KBs. One of the future challenges would be to ground into KBs in other languages and, in particular, in low resource languages. This setting is likely to introduce new sets of problems. The importance of dealing with new KBs comes up in other scenarios, ranging from the medical domain to various industrial domains, where corporations have a large number of KBs and they would like to ground mentions in their document to it. A common and yet important technical challenge in these settings is that of the "cold start problem", where new entities are mentioned in text, but they still don't have a representation that allows linking them. Better ways to deal with "cold start" is an important problem the future work will have to address.

End-to-End Training While the "standard" EDL pipeline consists of mention detection followed by an inference step, recent approaches have attempted to train such models in an end-to-end manner. These approaches might be useful in applications (e.g., (Li et al. 2020a)) but adapting it to the cross-lingual domain could be challenging, especially in the low-resource setting. On the other end, these settings could be useful when extending Linking beyond name entities, as suggested earlier, since the identification of mentions becomes an intermediate step in the process. The generalization abilities of these models, however, beyond their immediate application could suffer and more work is required to make sure that these approaches are not merely solving a specific data set.

References

Botha, J.A., Shan, Z., Gillick, D.: Entity Linking in 100 Languages. In: Proceedings of the 2020 Conference on Empirical Methods in Natural Language Processing (EMNLP), Association for Computational Linguistics, pp. 7833–7845. Online (2020). https://doi.org/10.18653/v1/2020.emnlp-main.630, https://aclanthology.org/2020.emnlp-main.630

Cao, N.D., Izacard, G., Riedel, S., Petroni, F.: Autoregressive entity retrieval. In: International Conference on Learning Representations (2021). https://openreview.net/forum?id=5k8F6UU39V

De Cao, N., Wu, L., Popat, K., Artetxe, M., Goyal, N., Plekhanov, M., Zettlemoyer, L., Cancedda, N., Riedel, S., Petroni, F.: Multilingual autoregressive entity linking. Trans. Assoc. Comput. Linguistics **10**, 274–290 (2022). https://doi.org/10.1162/tacl_a_00460. https://aclanthology.org/2022.tacl-1.16

Fu, X., Shi, W., Yu, X., Zhao, Z., Roth, D.: Design challenges in low-resource cross-lingual entity linking. In: Proceedings of the 2020 Conference on Empirical Methods in Natural Language Processing (EMNLP), Association for Computational Linguistics, pp. 6418–6432. Online (2020). https://doi.org/10.18653/v1/2020.emnlp-main.521, https://aclanthology.org/2020.emnlp-main.521

Gillick, D., Kulkarni, S., Lansing, L., Presta, A., Baldridge, J., Ie, E., Garcia-Olano, D.: Learning dense representations for entity retrieval. In: Proceedings of the 23rd Conference on Computational Natural Language Learning (CoNLL), Association for Computational Linguistics, pp. 528–537. Hong Kong, China (2019). https://doi.org/10.18653/v1/K19-1049, https://aclanthology.org/K19-1049

Hofstätter, S., Lin, S.-C., Yang, J.-H., Lin, J., Hanbury, A.: Efficiently teaching an effective dense retriever with balanced topic aware sampling. In: Proceedings of the 44th International ACM SIGIR Conference on Research and Development in Information Retrieval, pp. 113–122 (2021)

Johnson, J., Douze, M., Jégou, H.: Billion-scale similarity search with GPUs. IEEE Trans. Big Data (2019)

Li, B.Z., Min, S., Iyer, S., Mehdad, Y., Yih, W.-T.: Efficient one-pass end-to-end entity linking for questions. In: Proceedings of the 2020 Conference on Empirical Methods in Natural Language Processing (EMNLP), Association for Computational Linguistics, pp. 6433–6441. Online (2020a). https://doi.org/10.18653/v1/2020.emnlp-main.522, https://aclanthology.org/2020.emnlp-main.522

Reimers, N., Gurevych, I.: Sentence-BERT: Sentence embeddings using Siamese BERT-networks. In: Proceedings of the 2019 Conference on Empirical Methods in Natural Language Processing and the 9th International Joint Conference on Natural Language Processing (EMNLP-IJCNLP), Association for Computational Linguistics, pp. 3982–3992. Hong Kong, China (2019). https://doi.org/10.18653/v1/D19-1410, URL https://aclanthology.org/D19-1410

Wu, L., Petroni, F., Josifoski, M., Riedel, S., Zettlemoyer, L.: Scalable zero-shot entity linking with dense entity retrieval. In: Proceedings of the 2020 Conference on Empirical Methods in Natural Language Processing (EMNLP), Association for Computational Linguistics, pp. 6397–6407. Online (2020). https://doi.org/10.18653/v1/2020.emnlp-main.519, https://aclanthology.org/2020.emnlp-main.519

Appendix

A.1 Word Representations

In this section, we introduce basics of static word representations. Each word is represented as a vector of real numbers. These representations are static in the sense that they will not be changed according to the context in which they appear. Once the model is trained, the representations are fixed.

We will first discuss English word representations, including an earlier cluster-based word representation and a distributed word representation. The second part is about cross-lingual word representations, in which words from different languages are embedded in the same vector space. Therefore, two words of different languages can be compared directly using these representations.

A.1.1 Brown Clustering

Brown clustering (Brown et al. 1992) representation is a cluster-based word presentation, such that words that share meaningful linguistic properties gets assigned to the same cluster.

The input to the Brown clustering algorithm is a corpus of words $w_1, w_2, \ldots, w_T$, where T is the size of the corpus. The aim of the algorithm is to assign each word to a cluster, where $\mathcal{C} : V \rightarrow 1, 2, \ldots, K$ is a cluster assignment function that maps each word w_i in the vocabulary V to its cluster $\mathcal{C}(w_i)$. The sequence of cluster assignments $\mathcal{C}(w_1), \mathcal{C}(w_2), \ldots, \mathcal{C}(w_T)$ for words $w_1, w_2, \ldots, w_T$ is modeled using a *class-based bi-gram language model*. Under this model the generative story of the corpus is the following—transition to the current cluster $\mathcal{C}(w_i)$ conditioned on the previous cluster $\mathcal{C}(w_{i-1})$, sample a word w_i conditioned on the current cluster $\mathcal{C}(w_i)$, and repeat the process. The log-likelihood of the corpus $w_1, w_2, \ldots, w_T$ (or the *clustering quality*) under this model is computed as,

$$\log \Pr(w_1, w_2, \ldots, w_T) = \sum_{i=1}^{n} \log \Pr(w_i \mid \mathcal{C}(w_i)) \Pr(\mathcal{C}(w_i) \mid \mathcal{C}(w_{i-1})) \quad (A.1)$$

The output of the clustering algorithm is a binary tree, whose leaves are words, and internal nodes are clusters of the words in the sub-tree rooted at that node. Each cluster can be expressed using a binary code, which is the representation shared by all the words that belongs to that cluster. It can be shown that the clustering quality is a function of the mutual information of adjacent clusters,

$$Quality(\mathcal{C}) = \sum_{c,c'} \Pr(c, c') \log \frac{\Pr(c, c')}{\Pr(c)\Pr(c')} - \sum_{w} \Pr(w) \log \Pr(w) \quad (A.2)$$

$$= I(\mathcal{C}) - H \quad (A.3)$$

where c and c' are two clusters, $I(\mathcal{C})$ is the mutual information between adjacent clusters and H is the entropy of the word distribution.

Thus, Brown clustering uses mutual information of adjacent clusters (i.e., at bi-gram level) as a measure of distributional similarity between the words in those clusters. Starting with each word in its own distinct cluster, the brown clustering algorithm greedily merges pair of clusters that cause the smallest decrease in the likelihood of the corpus with respect to the class-based bi-gram language model. The greedy clustering strategy naturally leads to a hierarchical clustering upon termination.

Brown clustering has proved successful in several semi-supervised scenarios, improving performance on tasks like dependency parsing (Koo et al. 2008; Tratz and Hovy 2011), syntactic chunking (Turian et al. 2010), with sequence labeling tasks like named-entity recognition enjoying large gains (Freitag 2004; Miller et al. 2004; Ratinov and Roth 2009).

Limitations of Cluster-based Representations. A key limitation of using clustering-based representations is that they are not amenable to end-to-end learning aimed at a downstream task. This is due to the fact that the cluster assignment function $\mathcal{C}$ is, in general, discontinuous, so end-to-end learning methods like back-propagation cannot be used with them. Furthermore, approaches like Brown clustering scale *quadratically* in the number of clusters, and thus are extremely slow to train, preventing more expressive partitioning of the space. The discreteness of the space induced by the clustering also means that the granularity of similarity score between pair of words is discretely quantized. For instance, at a certain level of the Brown clusters, two words either belong to the same cluster (similarity of 1), or don't (similarity of 0). One can circumvent this by describing a word by its path from the root in the binary tree (e.g., *apple* becomes 101101). However, this approach partially alleviates the discreteness of the space.

A.1.2 Mono-Lingual Word Embedding

Vector-based word representations use a point in a n-dimensional vector space to describe each word in a language (see Figure A.1). Under this representation paradigm, geometric proximity in the vector space is used as surrogate for semantic similarity of two words. For instance, in Figure A.1, related words like *king* and *queen* are closer than *king* and *car*. The similarity of two words v and w represented in a vector space as $\mathbf{v}$ and $\mathbf{w}$ respectively, can then be computed using some natural geometric measure of vector proximity, like the dot product $dot(\mathbf{v}, \mathbf{w}) = \mathbf{v}^T \mathbf{w}$, or the cosine $cosine(\mathbf{v}, \mathbf{w}) = \frac{dot(\mathbf{v},\mathbf{w})}{\|\mathbf{v}\|\|\mathbf{w}\|}$. This notion of "proximity as a surrogate for similarity" traces its roots in cognition (Lakoff and Johnson 1980; Lakoff and Johnson 1999) as discussed in Sahlgren (2006).

As these representations *embed* a word in a geometric space, they are also called word *embeddings* or word *vectors*. Several word embedding models have been proposed in the literature. In the following, we introduce one of the most popular models: *skip-gram with negative sampling* or *word2vec*. For other models, please refer to the survey papers of word embeddings.

A.1.3 Skip-Gram with Negative Sampling (SGNS)

The language modeling objective only utilizes the past context (i.e., context to the left) to predict the next word. Mikolov et al. (2013c) proposed the *skip-gram model*, an alternative training strategy to learn word representations that utilizes both the right and left context around a word.

The skip-gram model's learning objective aims to train a classifier for a prediction task, and word representations are learnt as a by-product of this training objective. The prediction task aims to predict the surrounding words in a window around a central (or *pivot*) word,

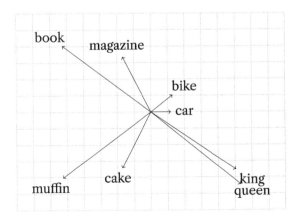

Fig. A.1 Vector-based word representations in a 2-dimensional space

using the representations of the pivot and surrounding words as parameters of the classifier. Formally, the learning objective of the skip-gram approach is,

$$\mathcal{L} = -\sum_{\mathbf{w}} \sum_{\mathbf{w}_c \in \text{NEIGHBORS}(\mathbf{w})} \log \Pr(\mathbf{w}_c \mid \mathbf{w})$$

The skip-gram formulation attempts to learn representations that are good at predicting neighboring words (in a predefined context window size), where the prediction are modeled using a log-linear conditional probability,

$$\Pr(\mathbf{w}_c \mid \mathbf{w}) = \frac{\exp(\mathbf{w}_c^T \mathbf{w})}{\sum_{\mathbf{w}' \in V} \exp(\mathbf{w}'^T \mathbf{w})} \quad (A.4)$$

The denominator of the conditional probability requires sum over $|V|$ terms, computing which can become prohibitive for vocabularies derived from large corpora. To avoid this computational bottleneck, efficient implementations of the skip-gram model employ *negative sampling* to optimize an approximation of the skip-gram objective. The *skip-gram with negative sampling (SGNS)* formulation poses the following binary classification task — is a given (w, c) pair a co-occurrence observed in the corpus D, or obtained by randomly pairing a word w with a context c? The randomly paired are referred as *negative samples* from a noise distribution D', leading the following learning objective,

$$\max \sum_{(w,c) \in D} \log \Pr(D = 1 \mid (w, c)) + \sum_{(w,c) \in D'} \log \Pr(D = 0 \mid (w, c)),$$

where r $\Pr(D = 1 \mid (w, c)) = \frac{1}{1+\exp(-\mathbf{w}^T \mathbf{c})} = 1 - \Pr(D = 0 \mid (w, c))$ is the probability that the model assigns to the (w, c) pair originating from the corpus. Note that the SGNS objective does not involve computing a normalization term over the entire vocabulary, and thus can be efficiently optimized. The negative sampling approach can be shown to be a special case of the more general technique of noise contrastive estimation (Gutmann and Hyvärinen 2012; Dyer 2014), used for training un-normalized probabilistic models. It was shown by Levy and Goldberg (2014) that the SGNS learning objective implicitly factorizes a shifted version of the PPMI matrix.

A.1.4 Cross-Lingual Word Embedding

Intuitively, if two words in English are geometrically close in the vector space, and one of them is missing its translation, the other word's nearest neighbor in the other language can serve as the missing translation (Fig. A.2).

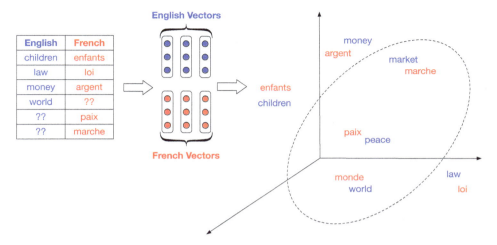

Fig. A.2 Cross-lingual word representations as a vector space approximation of discrete translation dictionaries. By encoding items present in the dictionary as points in a continuous vector space, one can recover some of the missing entries. Note that while the figure uses vector-based representations (which partition the space implicitly) a similar reasoning can be applied to cluster-based representations (which partition the space explicitly)

A.1.5 BiCCA

Bilingual Canonical Correlation Analysis based embeddings, henceforth referred as BiCCA, uses the projection technique described by Faruqui and Dyer (2014) to project independently trained monolingual embedding matrices $\mathbf{W}, \mathbf{V}$ using Canonical Correlation Analysis (CCA) (Hotelling 1936) to respect a bilingual dictionary. CCA is an algorithm that takes as input pairs of observations $\{(\mathbf{x}_i, \mathbf{y}_i)\}_{i=1}^{n}$ drawn from two different feature spaces ($\mathbf{x}_i \in \mathbf{X}, \mathbf{y}_i \in \mathbf{Y}$), and finds directions $\mathbf{d}_1$ and $\mathbf{d}_2$ such that the linear projections $\{\mathbf{d}_1^\mathsf{T}\mathbf{x}_i\}_{i=1}^{n}$ and $\{\mathbf{d}_2^\mathsf{T}\mathbf{y}_i\}_{i=1}^{n}$ are maximally correlated.

For applying CCA, first matrices $\mathbf{W}' \subseteq \mathbf{W}, \mathbf{V}' \subseteq \mathbf{V}$ are constructed such that $|\mathbf{W}'| = |\mathbf{V}'|$ and the corresponding rows ($\mathbf{w}_i, \mathbf{v}_i$) in the matrices are representations of words (w_i, v_i) that are translations of each other. The projection is then computed as:

$$\mathbf{P}_W, \mathbf{P}_V = \mathrm{CCA}(\mathbf{W}', \mathbf{V}') \quad (A.5)$$
$$\mathbf{W}^* = \mathbf{W}\mathbf{P}_W$$
$$\mathbf{V}^* = \mathbf{V}\mathbf{P}_V$$

Let l and m be the size of monolingual embeddings in $\mathbf{W}$ and $\mathbf{V}$ respectively, $\mathbf{P}_V \in \mathbb{R}^{l \times d}, \mathbf{P}_W \in \mathbb{R}^{m \times d}$ are the projection matrices with $d \leq \min(l, m)$ learned from CCA. Finally, the $\mathbf{V}^* \in \mathbb{R}^{|V| \times d}$ and $\mathbf{W}^* \in \mathbb{R}^{||W|| \times d}$ are the cross-lingual word embeddings of the two languages that are transformed from the correspondding monolingual embeddings.

Tsai and Roth (2016b) apply the BiCCA model on Wikipedia articles to generate cross-lingual word and entity embeddings for cross-lingual entity linking. They first train monolingual embeddings for each language separately using the "Alignment by Wikipedia Anchors" idea proposed in (Wang et al. 2014). For each language, they take all documents in Wikipedia and replace the hyperlinked text with the corresponding Wikipedia title. For example, consider the following Wikipedia sentence: "It is led by and mainly composed of **Sunni** Arabs from **Iraq** and **Syria**.", where the three bold faced mentions are linked to some Wikipedia titles. After the applying the proposed transformation, the sentence becomes "It is led by and mainly composed of `Sunni_Islam` Arabs from `en/Iraq` and `en/Syria`." The skip-gram model (Mikolov et al. 2013a; Mikolov et al. 2013c) is then applied on this newly generated text to produce monolingual embeddings. Since Wikipedia titles appear as tokens in the transformed text, both words and entities are embedded in the same vector space.

Next, they apply the BiCCA model on these monolingual embeddings. The requirement of this model is a dictionary which maps the words in one language to the words in another language (the $\mathbf{W}'$ and $\mathbf{V}'$ in Eq. A.5). Note that there is no need to have this mapping for every word in the vocabularies. The aligned words are only used to learn the projection matrices, which can later be applied on embeddings of all words. Faruqui and Dyer (2014) obtain this dictionary by picking the most frequent translated word from a parallel corpus. However, there is a limited or no parallel corpus for many languages. Since the proposed monolingual embedding model consists also of Wikipedia title embeddings, the alignments between Wikipedia titles via inter-language links can be used to construct $\mathbf{W}'$ and $\mathbf{V}'$ in Eq. A.5, which is the key idea in Tsai and Roth (2016b).

The principle in the BiCCA model is the basis of many representation learning algorithms that use bilingual' dictionaries (Mikolov et al. 2013b; Lu et al. 2015; Smith et al. 2017; Artetxe et al. 2017).

A.2 Context Representations

A.2.1 RNN and LSTM

Recurrent neural networks (RNNs) are a family of neural networks that take as input a sequence of vectors $(\mathbf{x}_1, \mathbf{x}_2, \ldots, \mathbf{x}_n)$ and return another sequence $(\mathbf{h}_1, \mathbf{h}_2, \ldots, \mathbf{h}_n)$ that represents some information about the sequence at every step in the input. More precisely, RNNs apply a parameterized function recursively on a sequence $(\mathbf{x}_1, \mathbf{x}_2, \ldots, \mathbf{x}_n)$:

$$\mathbf{h_t} = f(\mathbf{h_{t-1}}, \mathbf{x_t}; \Theta),$$

where the function f is parameterized by Θ. For NLP applications, a sequence is usually a sentence, so a vector $\mathbf{x}_i$ in the sequence represents the i-th word in the sentence. The word representations that we discussed in the previous sections could be used to initialize these $\mathbf{x}$ vectors.

Long Short-term Memory Networks (LSTMs) are a special class of RNNs that incorporate a memory-cell and several gates that control the proportion of the input to give to the memory cell, and the proportion from the previous state to forget to capture long-range dependencies (Hochreiter and Schmidhuber 1997). Mathmatically, LSTMs can be formulated as follows:

$$\mathbf{f}_t = \sigma(W_f \mathbf{x}_t + U_f \mathbf{h}_{t-1} + \mathbf{b}_f)$$
$$\mathbf{i}_t = \sigma(W_i \mathbf{x}_t + U_i \mathbf{h}_{t-1} + \mathbf{b}_i)$$
$$\mathbf{o}_t = \sigma(W_o \mathbf{x}_t + U_o \mathbf{h}_{t-1} + \mathbf{b}_o)$$
$$\mathbf{g}_t = \tanh(W_g \mathbf{x}_t + U_g \mathbf{h}_{t-1} + \mathbf{b}_g)$$
$$\mathbf{c}_t = \mathbf{f}_t \odot \mathbf{c}_{t-1} + \mathbf{i}_t \odot \mathbf{g}_t$$
$$\mathbf{h}_t = \mathbf{o}_t \odot \tanh(\mathbf{c}_t),$$

where σ is the element-wise sigmoid function, and $\odot$ is the element-wise product. If $\mathbf{x} \in R^d$ and $\mathbf{h} \in R^h$, $W_f, W_i, W_o, W_g \in R^{h \times d}$, $U_f, U_i, U_o, U_g \in R^{h \times h}$, and $\mathbf{b}_f, \mathbf{b}_i, \mathbf{b}_o, \mathbf{b}_g \in R^h$ are the model parameters to be learned.

For a given sentence $(\mathbf{x}_1, \mathbf{x}_2, \ldots, \mathbf{x}_n)$ containing n words, an LSTM representation at the t-th position $\mathbf{h}_t$ encodes information from the beginning of the sentence up to word $\mathbf{x}_t$. In order to also incorporate information to the right of the word, a natural extension is to run another backward LSTM which encodes the sentence from right to left instead. The hidden representation generated from the forward pass is usually denoted as $\overrightarrow{\mathbf{h}_t}$ and the one from the backward LSTM is $\overleftarrow{\mathbf{h}_t}$. The final context representation at word $\mathbf{x}_t$ is then defined as $\mathbf{h}_t = [\overrightarrow{\mathbf{h}_t}; \overleftarrow{\mathbf{h}_t}]$, the concatenation of the hidden vectors from both directions. This model is usually referred as Bidirectional LSTM or BiLSTM. The BiLSTM model is widely used as the context encoder in NER models.

A.2.2 Pre-trained Context Representations

BERT. Bidirectional Encoder Representations from Transformers (BERT) are pre-trained contexual representations, obtained by training a transformer model on large amounts of unlabeled text (Devlin et al. 2019). When training BERT representations, the transformer takes as input two sentences A and B, and optimizes for two learning objectives. A fraction of the words in each sentence is replaced by a token [MASK], denoting this word's identity has been masked. The first objective, referred to as the the Masked Language Modeling (MLM) objective, aims to predict the masked words in a sentence. At the same time, the transformer is trained to predict if two sentences appear next to each other in a corpus, corresponding to the next sentence prediction (NSP) objective. Figure A.3 shows the two training tasks the transformer optimizes for in BERT pre-training.

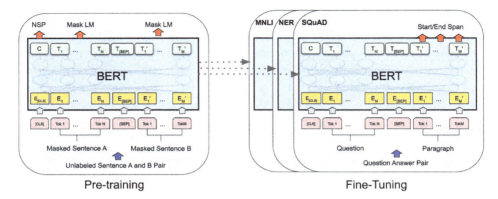

Fig. A.3 BERT pre-training and fine-tuning frameworks from Devlin et al. (2019). When pre-training, BERT takes as input two sentences A and B. The pre-trained model is then fine-tuned for downstream tasks

The pre-trained representations obtained thus can be used in several downstream tasks. The most common approach involves fine-tuning the representations, in which a new layer is added on top of the BERT representations, and trained for the task specific objective.
m-BERT. m-BERT is the multilingual version of BERT, where the input to the transformer during training consists of sentence pairs from multiple languages. Specifically, m-BERT is trained using Wikipedia text from 104 languages, all of which share a common word-piece vocabulary. Some studies on understanding and analyzing m-BERT include Karthikeyan et al. (2019), Wu and Dredze (2019), and Pires et al. (2019).

A.3 Machine Learning Models

In this section, we introduce common machine learning models which have been used to solve either named entity recognition or entity linking problem. All these models are supervised learning models which require labeled training data for learning model parameters. We will discuss two binary classification models: perceptron and SVMs, a ranking model (ranking SVM), a structured prediction model (CRF), and an ensemble technique (gradient boosting) which combines predictions from several weak models.

A.3.1 Perceptron

Perceptron is an algorithm for learning a linear binary classifier. At inference time, an instance $\mathbf{x} \in R^n$ is scored by a linear function

$$f(\mathbf{x}) = \mathbf{w}^T \mathbf{x} + b, \tag{A.6}$$

where $\mathbf{w} \in R^n$ and b are model parameters. The prediction $\hat{y}$ can be computed as follow

$$\hat{y} = \begin{cases} +1 & \text{if } f(\mathbf{x}) \geq 0, \\ -1 & \text{if } f(\mathbf{x}) < 0 \end{cases} \quad (A.7)$$

Given training data $(\mathbf{x}_i, y_i), i = 1, \ldots, m$, $\mathbf{x_i} \in R^n$ are feature vectors and $y \in \{+1, -1\}$ are binary labels, several algorithms have been proposed for finding $\mathbf{w}$ in Eq. (A.6) based on different learning objective. Perceptron (Rosenbaltt 1957) is one of the earliest approach. Following is the Perceptron learning rules.

1. Initialize model parameters $\mathbf{w}$ and b.
2. Cycle through training examples multiple times. For each example $(\mathbf{x}_i, y_i)$,

    a. Compute predicted label $\hat{y}_i$ by Eq. (A.7)
    b. If $\hat{y}_i \neq y_i$, update the weight vector: $\mathbf{w} \leftarrow \mathbf{w} + r y_i \mathbf{x}_i$, where r is a constant learning rate.
    Otherwise, if $\hat{y}_i = y_i$, leave weights unchanged.

This learning process is guaranteed to converge if the training data is linearly separable. Usually, a fixed number of rounds to go through the entire training data is specified as the stopping criteria. We note that several variants of Perceptron have been proposed, for instance, the Maxover algorithm (Wendemuth 1995) is more robust in terms of convergence when the training data is not linearly separable. A much more practical alternative is the Averaged Perceptron. The key idea is that Averaged Perceptron returns a weighted average of earlier hypotheses, and the weights are a function of the length of no-mistakes stretch. Averaged Perceptron has been successfully applied on the NER problem as we discussed in Chap. 4.

A.3.2 Support Vector Machines

Similar to the linear classification idea of perceptron, support vector machines (SVMs) are supervised binary classification models with associated learning algorithms that find $\mathbf{w}$ and b in Eq. (A.6) and use Eq. (A.7) to compute predictions. The key difference from perceptron is that SVMs obtain model parameters $\mathbf{w}$ and b via optimizing a loss function. Given training data $(\mathbf{x}_i, y_i), i = 1, \ldots, m$, $\mathbf{x_i} \in R^n$ are feature vectors and $y \in \{+1, -1\}$ are binary labels, Boser et al. (1992), Cortes and Vapnik (1995) solved the following primal optimization problem:

$$\min_{\mathbf{w},b,\xi} \quad \frac{1}{2}\mathbf{w}^T\mathbf{w} + C\sum_{i=1}^{m}\xi_i$$

$$\text{subject to} \quad y_i(\mathbf{w}^T\phi(\mathbf{x}_i)+b) \geq 1-\xi_i$$

$$\xi_i \geq 0, i=1,\ldots,m,$$

where ϕ is a function that maps a feature vector $\mathbf{x}$ into a higher-dimensional space, and C is the regularization hyper-parameter. Because of this ϕ mapping function, the model parameter $\mathbf{w}$ could be in a very high dimensional space. In this case, people usually solve the dual problem instead:

$$\min \quad \frac{1}{2}^T Q - \mathbf{e}^T \qquad (A.8)$$

$$\text{subject to} \quad \mathbf{y}^T = 0,$$

$$0 \leq \alpha_i \leq C, i=1,\ldots,m,$$

where $\mathbf{e}$ is a vector of all ones, Q is an m by m positive semidefinite matrix, $Q_{ij} = y_i y_j K(\mathbf{x_i}, \mathbf{x_j})$, and $K(\mathbf{x_i}, \mathbf{x_j}) = \phi(\mathbf{x_i})^T \phi(\mathbf{x_j})$ is the kernel function.

After Eq. (A.8) is solved, the optimal $\mathbf{w}$ can be computed based on the primal-dual relationship:

$$\mathbf{w} = \sum_{i=1}^{m} y_i \alpha_i \phi(\mathbf{x}_i)$$

and hence the decision function Eq. (A.7) becomes

$$\text{sgn}(\mathbf{w}^T \phi(\mathbf{x}) + b) = \text{sgn}(\sum_{i=1}^{m} y_i \alpha_i K(\mathbf{x}_i, \mathbf{x}) + b).$$

When the kernel function is not used, that is, $\phi(\mathbf{x}) = \mathbf{x}$, the model is usually referred as *linear* SVMs. Otherwise, it is called *non-linear* or *kernel* SVMs, since a linear function learnt in a higher-dimensional feature space is usually non-linear in the original feature space. For NLP problems in which the feature vectors are usually pretty high-dimensional (e.g., n-gram features), a linear SVM could already achieve good performance.

SVMs were very popular for any problem that can be formulated as a classification problem. They have been applied to solve the NER problem (Chap. 4) and to learn the scoring function for ranking candidate entities (Chap. 6).

A.3.3 RankSVM

In the learning to rank setup, we are given a set of training label-query-instance tuples $(y_i, q_i, \mathbf{x}_i)$, where $y_i \in R$, $\mathbf{x}_i \in R^n$, $i=1,\ldots,m$, and q_i is a query ID. A query q_i is usually associated with multiple instances. Given a query, $y_i > y_j$ means that $\mathbf{x}_i$ is more relevant than $\mathbf{x}_j$. By defining the set of preference pairs as

$$P \equiv \{(i, j) \mid q_i = q_j, y_i > y_j\},$$

rankSVM (Herbrich et al. 2000; Joachims 2002) solves

$$\min_{\mathbf{w}, \cdot} \frac{1}{2}\mathbf{w}^T\mathbf{w} + C \sum_{(i,j) \in P} \xi_{i,j} \qquad (A.9)$$

$$\text{subject to} \quad \mathbf{w}^T(\phi(\mathbf{x}_i) - \phi(\mathbf{x}_j)) \geq 1 - \xi_{i,j}$$

$$\xi_{i,j} \geq 0, \forall (i, j) \in P,$$

where $C > 0$ is the regularization hyper-parameter and ϕ is a function maps features to a higher-dimensional space. The summation of $\xi_{i,j}$ in the second term of Eq. (A.9) is called L1 loss. Another common choice is L2 loss (replace $\xi_{i,j}$ by $\xi_{i,j}^2$).

The key idea behind rankSVM is to learn a $\mathbf{w}$ such that $\mathbf{w}^T\phi(\mathbf{x}_i) > \mathbf{w}^T\phi(\mathbf{x}_j)$ if $(i, j) \in P$. Therefore, rankSVM is considered as a pairwise ranking model. We can see that the relevance scores, y, do not show up in the objective function. They only affect the selection of pairs of instances in set P. In other words, the magnitude of relevance scores does not matter. All it cares about is the order of instances. Similar to the SVMs classification model, the kernel tricks on the dual form can also be applied here.

At inference time, a test instance $\mathbf{x}$ is simply scored by $\mathbf{w}^T\mathbf{x}$. The higher the score, the more important the instance is.

A.3.4 Linear-Chain Conditional Random Fields

The Perceptron and SVMs classifiers introduced in the previous sections assume instances are independent. Making prediction on $\mathbf{x}_i$ does not depend on another instance $\mathbf{x}_j$ and its label y_j. However, this assumption does not hold for many NLP applications. As we discussed in Chap. 4, NER is usually formulated as a word classification problem, in which a model would predict a label for each word. These labels can then be used to decode which consecutive words should form a named entity mention. In this setup, the decision on one word is clearly dependent on other words, especially the neighboring words. For instance, if a word is classified as "B-LOC" (beginning of a location phrase), the next word is unlikely to be labeled as "I-PER" (inside a person phrase).

In order to relax the independence assumption, one would need to make inference on all words in a sentence simultaneously. However, the number of possible label combinations for a sentence could be very large, making inference intractable. People usually make dependency assumption among words to simplify the problem, since the label of a word usually does not depend on labels of every other words in the sentence. The simplest and the most common dependency structure in NLP is the *linear-chain* structure. Words in a sentence are arranged in a linear chain. The label y_i of a word x_i only depends on the label of previous word y_{i-1}.

Let $x_{1:N}$ be the sequence of words in a sentence, and $y_{1:N}$ be the corresponding labels. A linear-chain Conditional Random Field defines a conditional probability:

$$p(y_{1:N}|x_{1:N}) = \frac{1}{Z}\exp\left(\sum_{i=1}^{N}\sum_{k=1}^{K} w_k f_k(y_{i-1}, y_i, x_{1:N})\right), \tag{A.10}$$

where the scalar Z is the normalization factor, or partition function, which makes $p(y_{1:N}|x_{1:N})$ a valid probability. That is, Z sums over all possible label sequences $y_{1:N}$,

$$Z = \sum_{y_{1:N}} \exp\left(\sum_{i=1}^{N}\sum_{k=1}^{K} w_k f_k(y_{i-1}, y_i, x_{1:N})\right).$$

Since the number of possible label sequences is exponentially large, Z is difficult to compute in general. However, under the linear-chain assumption, it can be computed efficiently by the forward-backward algorithm. We omit the details here and refer the readers to articles which focus on CRF (Sutton and McCallum 2006).

Inside the big exponential function of Eq. (A.10), the label score of each word position i is a linear combination of model parameters w_k and feature functions f_k. The feature functions are the key components of CRF. In the case of linear-chain CRF, the general form of feature functions is $f_k(y_{i-1}, y_i, x_{1:N})$, which depends on y_i (the label for the current word x_i), y_{i-1} (the label for the previous word x_{i-1}), and $x_{1:N}$ (words in the entire sequence). These functions could be arbitrary functions that output a real value. For instance, we could have a simple binary feature which only looks at the current word and label:

$$f_1(y_{i-1}, y_i, x_{1:N}) = \begin{cases} 1 & \text{if } y_i = \text{B-PER and } x_i = \text{Alex}, \\ 0 & \text{otherwise}. \end{cases}$$

To label an unseen sequence at inference time, the most likely labelling $y_{1:N}^* = \arg\max_{y_{1:N}} p(y_{1:N}|x_{1:N})$ can be computed efficiently and exactly using the Viterbi algorithm, a dynamic-programming technique.

Besides the NER problem we discussed in Chap. 4, linear-chain CRF is a very common model for sequence labelling tasks in NLP, such as Word Segmentation, POS Tagging, Mention Detection, Relation Extraction.

A.4 Inference Methods

In Chap. 5, we discussed global inference for entity linking problem. Since global inference formalism makes decision for multiple mentions simultaneously, the problem may not be solved efficiently in general. In this section, we introduce two inference techniques that have been applied to disambiguate entity mentions jointly: integer linear programming and belief propagation.

A.4.1 Integer Linear Programming

In an integer linear programming (ILP) problem, we seek to maximize a linear cost function over all n-dimensional vectors $\mathbf{x}$ subject to a set of linear inequality constraints:

$$\max \quad \mathbf{c}^T \mathbf{x} \quad \text{(A.11)}$$
$$\text{subject to} \quad A\mathbf{x} \leq \mathbf{b},$$
$$\mathbf{x} \geq 0,$$
$$\mathbf{x} \in \mathbb{Z}^n,$$

where $\mathbf{c}$ and $\mathbf{b}$ are two given n-dimensional vectors, A is a matrix, and the vector $\mathbf{x}$ contains only integers.

Many combinatorial optimization problems can be formulated as an ILP. In the context of classification, when instances are interdependent and we want to jointly label several instances together, this MAP inference problem could usually be formulated as an ILP problem. For example, as we discussed in Sect. A.3.4 (CRF), the linear-chain assumption makes labelling a sequence of words efficiently. However, when the dependency structure among words is more complex, formulating the inference problem as an ILP may help to find a good solution more quickly.

In addition, ILP formulations allow us to naturally introduce constraints. One can incorporate additional domain knowledge during the inference time which might be hard to be learnt from the data during training time. For instance, taking POS tagging as an example, one could introduce a constraint such as "there must be at least one verb in a sentence" via the linear inequality in Eq. (A.11).

In the Entity Linking context, Cheng and Roth (2013) jointly link several mentions to candidate entities using ILP formulation:

$$\arg\max_{e,r} \quad \sum_i \sum_k s_i^k e_i^k + \sum_{i,j} \sum_{k,l} w_{i,j}^{(k,l)} r_{i,j}^{(k,l)}$$

$$\text{subject to} \quad e_i^k \in \{0, 1\}, \quad \text{(Integral constraints)}$$

$$r_{i,j}^{(k,l)} \in \{0, 1\}, \quad \text{(Integral constraints)}$$

$$\sum_k e_i^k = 1, \quad \text{(Unique solution)}$$

$$2r_{i,j}^{(k,l)} \leq e_i^k + e_j^l, \quad \text{(Relation definition)}$$

where i and j are mention indices, and k and l are indices for candidate entities. The boolean variable e_i^k indicates if mention i is linked to its k-th candidate entity, and s_i^k is some local ranking score for this candidate entity. In addition, a mention can only be linked to exactly one of its candidate entities, as specified in the unique solution constraints. Another boolean variable $r_{i,j}^{(k,l)}$ indicates if mention i is linked to its k-th candidate and mention j is linked to its l-th candidate. The relational score between these two entities is given as $w_{i,j}^{(k,l)}$. The last set of constraints specify the relationship between these two types of boolean variables. For instance, if $r_{i,j}^{(k,l)}$ is one, then both e_i^k and e_j^l should also be one. We can see that this ILP formulation nicely utilize both local and relational scores, and makes inference for multiple mentions jointly.

Solving ILP is NP hard in general. A naive way to solve an ILP is to relax the integral constraint on **x**, solve the corresponding linear programming (LP) problem, and then round the solution to integers. However, this solution may not be optimal or may even violate some constraints. Nevertheless, if A and **b** have all integer entries and A is totally unimodular, solution from the LP relaxation is guaranteed to be integral. For the general case, a variety of algorithms based on LP relaxation have been proposed to solve ILP, such as cutting plane methods and branch and bound methods. Off-the-shelf ILP solvers are quite efficient and robust (e.g., Gurobi[1] and CPLEX[2]), which make ILP an attractive formalism for solving inference problems in NLP.

A.4.2 Belief Propagation

Belief Propagation is a message passing algorithm for performing inference on probabilistic graphical models. It can be used to compute the marginal distribution for each unobserved variables, or to compute the maximum a posteriori probability (MAP) of unobserved variables.

---

[1] https://www.gurobi.com/.
[2] https://www.ibm.com/analytics/cplex-optimizer.

Fig. A.4 An undirected graphical model with observed variables y_1, y_2, and y_3 and unobserved variables x_1, x_2, and x_3

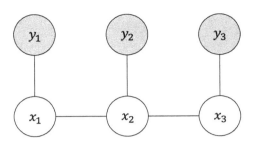

Let's consider a simple Markov chain model with three observed and three unobserved variables as shown in Fig. A.4. According to the dependency structure, its joint probability is in the form of:

$$P(x_1, x_2, x_3, y_1, y_2, y_3) = \phi_{12}(x_1, x_2)\phi_{23}(x_2, x_3)\psi_1(x_1, y_1)\psi_2(x_2, y_2)\psi_3(x_3, y_3).$$

The marginal probability of x_1:

$$P(x_1|\mathbf{y}) = \frac{1}{P(\mathbf{y})} \sum_{x_2} \sum_{x_3} P(x_1, x_2, x_3, y_1, y_2, y_3) \quad (A.12)$$

$$= \frac{1}{P(\mathbf{y})} \sum_{x_2} \sum_{x_3} \phi_{12}(x_1, x_2)\phi_{23}(x_2, x_3)\psi_1(x_1, y_1)\psi_2(x_2, y_2)\psi_3(x_3, y_3) \quad (A.13)$$

$$= \frac{1}{P(\mathbf{y})} \psi_1(x_1, y_1) \sum_{x_2} \phi_{12}(x_1, x_2)\psi_2(x_2, y_2) \sum_{x_3} \phi_{23}(x_2, x_3)\psi_3(x_3, y_3) \quad (A.14)$$

$$= \frac{1}{P(\mathbf{y})} \psi_1(x_1, y_1) \sum_{x_2} \phi_{12}(x_1, x_2)\psi_2(x_2, y_2) \sum_{x_3} \phi_{23}(x_2, x_3) m_{y_3 \to x_3}(x_3)$$
$$(A.15)$$

$$= \frac{1}{P(\mathbf{y})} \psi_1(x_1, y_1) \sum_{x_2} \phi_{12}(x_1, x_2)\psi_2(x_2, y_2) m_{x_3 \to x_2}(x_2) \quad (A.16)$$

$$= \frac{1}{P(\mathbf{y})} \psi_1(x_1, y_1) m_{x_2 \to x_1}(x_1) \quad (A.17)$$

$$= \frac{1}{P(\mathbf{y})} m_{y_1 \to x_1}(x_1) m_{x_2 \to x_1}(x_1), \quad (A.18)$$

where $m_{y_3 \to x_3}(x_3)$ denotes the message passed from y_3 to x_3. We can see that a message is simply a partial sum in this marginal calculation. The derivation from Eqs. (A.15)–(A.18) depicts the process of message passing, which starts from y_3 passes a message to x_3, and then from x_3 to x_2, and finally from x_2 to the target variable x_1. Factorizing the double sum in Eq. (A.14) could largely reduce computational cost. In addition, these messages are reusable in computing marginal probabilities for other nodes. For instance, when computing the marginal probability of x_2,

$$P(x_2|\mathbf{y}) = \frac{1}{P(\mathbf{y})} \psi_2(x_2, y_2) \sum_{x_1} \phi_{12}(x_1, x_2) \psi_1(x_1, y_1) \sum_{x_3} \phi_{23}(x_2, x_3) \psi_3(x_3, y_3),$$

we can see that the last summation over x_3 is identical to the one in Eq. (A.14). In other words, messages are re-usable partial sums in the marginal calculations.

In general, for Markov Random Fields, a message from node x_j to node x_i can be written as:

$$m_{x_j \to x_i}(x_i) = \sum_{x_j} \psi_{ij}(x_i, x_j) \prod_{x_k \in n(x_j) \setminus x_i} m_{x_k \to x_j}(x_j),$$

where $n(x_j) \setminus x_i$ is the set of neighboring nodes of x_j except x_i. The marginal probability of node x_i is the product of all incoming messages:

$$P(x_i) = \prod_{x_j \in n(x_i)} m_{x_j \to x_i}(x_i).$$

This algorithm of computing marginal probabilities is also called *sum-product* algorithm since a message is a sum of products. Instead of summing over the states of other unobserved variables, sometimes we are interested in finding the most probable states of all unobserved variables. In this case, the same algorithm works by replacing summation with arg max operator. This becomes the *max-product* version of belief propagation.

The above belief propagation algorithm can give the exact solution when the graph is a tree or a chain. For general graphs with loops, it was found that belief propagation can still be used to obtain a good approximate solution. The modified algorithm is sometimes called *loopy belief propagation*. We omit the details in this book, and refer the interested readers to more dedicated articles such as Ihler et al. (2005) and Murphy et al. (2013).

References

Agić, Ž., Hovy, D., Søgaard, A.: If all you have is a bit of the Bible: Learning POS taggers for truly low-resource languages. In: Proceedings of the 53rd Annual Meeting of the Association for Computational Linguistics and the 7th International Joint Conference on Natural Language Processing (Volume 2: Short Papers), Association for Computational Linguistics, pp. 268–272. Beijing, China (2015). https://doi.org/10.3115/v1/P15-2044, https://aclanthology.org/P15-2044

Agić, Ž., Johannsen, A., Plank, B., Martínez Alonso, H., Schluter, N., Søgaard, A.: Multilingual projection for parsing truly low-resource languages. Trans. Assoc. Comput. Linguistics **4**, 301–312 (2016). https://doi.org/10.1162/tacl_a_00100, https://aclanthology.org/Q16-1022

Akbik, A., Blythe, D., Vollgraf, R.: Contextual string embeddings for sequence labeling. In: Proceedings of the 27th International Conference on Computational Linguistics, Association for Computational Linguistics, pp. 1638–1649. Santa Fe, New Mexico, USA (2018). https://aclanthology.org/C18-1139

Ammar, W., Dyer, C., Smith, N.: Transliteration by sequence labeling with lattice encodings and reranking. In: Proceedings of the 4th Named Entity Workshop (NEWS) 2012, Association for Computational Linguistics, pp. 66–70. Jeju, Korea (2012). https://aclanthology.org/W12-4410

Arora, R., Tsai, C.-T., Tsereteli, K., Kambadur, P., Yang, Y.: A semi-Markov structured support vector machine model for high-precision named entity recognition. In: Proceedings of the 57th Annual Meeting of the Association for Computational Linguistics, Association for Computational Linguistics, pp. 5862–5866. Florence, Italy (2019). https://doi.org/10.18653/v1/P19-1587, https://aclanthology.org/P19-1587

Artetxe, M., Labaka, G., Agirre, E.: Learning bilingual word embeddings with (almost) no bilingual data. In: Proceedings of the 55th Annual Meeting of the Association for Computational Linguistics (Volume 1: Long Papers), Association for Computational Linguistics, pp. 451–462. Vancouver, Canada (2017). https://doi.org/10.18653/v1/P17-1042, https://aclanthology.org/P17-1042

Auer, S., Bizer, C., Kobilarov, G., Lehmann, J., Cyganiak, R., Ives, Z.: DBpedia: a nucleus for a web of open data. The Semantic Web 722–735 (2007)

Bada, M., Eckert, M., Evans, D., Garcia, K., Shipley, K., Sitnikov, D., Baumgartner, W.A., Cohen, K.B., Verspoor, K., Blake, J.A., Hunter, L.E.: Concept annotation in the CRAFT corpus. BMC Bioinform. (2012)

Bagga, A., Baldwin, B.: Entity-based cross-document coreferencing using the vector space model. In: COLING 1998 Volume 1: The 17th International Conference on Computational Linguistics (1998). https://aclanthology.org/C98-1012

Balasuriya, D., Ringland, N., Nothman, J., Murphy, T., Curran, J.R.: Named entity recognition in Wikipedia. In: Proceedings of the 2009 Workshop on The People's Web Meets NLP: Collaboratively Constructed Semantic Resources (People's Web), Association for Computational Linguistics, pp. 10–18. Suntec, Singapore (2009). https://aclanthology.org/W09-3302

Banarescu, L., Bonial, C., Cai, S., Georgescu, M., Griffitt, K., Hermjakob, U., Knight, K., Koehn, P., Palmer, M., Schneider, N.: Abstract meaning representation for sembanking. In: Proceedings of the 7th Linguistic Annotation Workshop and Interoperability with Discourse, Association for Computational Linguistics, pp. 178–186. Sofia, Bulgaria (2013). https://aclanthology.org/W13-2322

Basaldella, M., Liu, F., Shareghi, E., Collier, N.: COMETA: a corpus for medical entity linking in the social media. In: Proceedings of the 2020 Conference on Empirical Methods in Natural Language Processing (EMNLP), Association for Computational Linguistics, pp. 3122–3137. Online (2020). https://doi.org/10.18653/v1/2020.emnlp-main.253, https://aclanthology.org/2020.emnlp-main.253

Bodenreider, O.: The unified medical language system (UMLS): integrating biomedical terminology. Nucleic Acids Res (2004)

Bollacker, K., Evans, C., Paritosh, P., Sturge, T., Taylor, J.: Freebase: a collaboratively created graph database for structuring human knowledge. In: Proceedings of the 2008 ACM SIGMOD international conference on Management of data. ACM (2008)

Boser, B.E., Guyon, I.M., Vapnik, V.N.: A training algorithm for optimal margin classifiers. In: Proceedings of the Fifth Annual Workshop on Computational Learning Theory, pp. 144–152 (1992)

Botha, J.A., Shan, Z., Gillick, D.: Entity Linking in 100 Languages. In: Proceedings of the 2020 Conference on Empirical Methods in Natural Language Processing (EMNLP), Association for Computational Linguistics, pp. 7833–7845. Online (2020). https://doi.org/10.18653/v1/2020.emnlp-main.630, https://aclanthology.org/2020.emnlp-main.630

Brown, P.F., Della Pietra, V.J., deSouza, P.V., Lai, J.C., Mercer, R.L.: Class-based n-gram models of natural language. Comput. Linguistics **18**(4), 467–480 (1992). https://aclanthology.org/J92-4003

Bunescu, R., Paşca, M.: Using encyclopedic knowledge for named entity disambiguation. In: 11th Conference of the European Chapter of the Association for Computational Linguistics, pp. 9–16. Trento, Italy (2006). https://aclanthology.org/E06-1002

Burger, J.D., Henderson, J.C., Morgan, W.T.: Statistical named entity recognizer adaptation. In: COLING-02: The 6th Conference on Natural Language Learning 2002 (CoNLL-2002) (2002). https://aclanthology.org/W02-2003

Cai, J., Strube, M.: Evaluation metrics for end-to-end coreference resolution systems. In: Proceedings of the SIGDIAL 2010 Conference, Association for Computational Linguistics, pp. 28–36. Tokyo, Japan (2010). https://aclanthology.org/W10-4305

Cao, N.D., Izacard, G., Riedel, S., Petroni, F.: Autoregressive entity retrieval. In: International Conference on Learning Representations (2021). https://openreview.net/forum?id=5k8F6UU39V

Cao, Z., Qin, T., Liu, T., Tsai, M., Li, H.: Learning to rank: from pairwise approach to listwise approach. In: Ghahramani, Z. (ed.) Proceedings of the International Conference on Machine Learning (ICML) (2007)

References

Carreras, X., Màrquez, L., Padró, L.: Named entity extraction using AdaBoost. In: COLING-02: The 6th Conference on Natural Language Learning 2002 (CoNLL-2002) (2002). https://aclanthology.org/W02-2004

Chabchoub, M., Gagnon, M., Zouaq, A.: FICLONE: improving DBpedia spotlight using named entity recognition and collective disambiguation. Open J. Semantic Web (OJSW) **5**(1), 12–28 (2018)

Chang, M.-W., Goldwasser, D., Roth, D., Tu, Y.: Unsupervised constraint driven learning for transliteration discovery. In: Proceedings of Human Language Technologies: The 2009 Annual Conference of the North American Chapter of the Association for Computational Linguistics, Association for Computational Linguistics, pp. 299–307. Boulder, Colorado (2009). https://aclanthology.org/N09-1034

Chang, M.-W., Ratinov, L., Roth, D.: Structured learning with constrained conditional models. Machine Learn. **88**(3), 399–431 (2012). http://cogcomp.org/papers/ChangRaRo12.pdf

Chen, G., Ma, S., Chen, Y., Zhang, D., Pan, J., Wang, W., Wei, F.: Towards making the most of cross-lingual transfer for zero-shot neural machine translation. In: Proceedings of the 60th Annual Meeting of the Association for Computational Linguistics (Volume 1: Long Papers), pp. 142–157. Dublin, Ireland (2022). https://doi.org/10.18653/v1/2022.acl-long.12, https://aclanthology.org/2022.acl-long.12

Chen, H., Li, X., Zukov Gregoric, A., Wadhwa, S.: Contextualized end-to-end neural entity linking. In: Proceedings of the 1st Conference of the Asia-Pacific Chapter of the Association for Computational Linguistics and the 10th International Joint Conference on Natural Language Processing, Association for Computational Linguistics, pp. 637–642. Suzhou, China (2020). https://aclanthology.org/2020.aacl-main.64

Chen, Y.-H., Choi, J.D.: Character identification on multiparty conversation: Identifying mentions of characters in TV shows. In: Proceedings of the 17th Annual Meeting of the Special Interest Group on Discourse and Dialogue, Association for Computational Linguistics, pp. 90–100. Los Angeles (2016). https://doi.org/10.18653/v1/W16-3612, https://aclanthology.org/W16-3612

Chen, Z., Ji, H.: Collaborative ranking: a case study on entity linking. In: Proceedings of the 2011 Conference on Empirical Methods in Natural Language Processing, Association for Computational Linguistics, pp. 771–781. Edinburgh, Scotland, UK (2011). https://aclanthology.org/D11-1071

Cheng, X., Roth, D.: Relational inference for wikification. In: Proceedings of the 2013 Conference on Empirical Methods in Natural Language Processing, Association for Computational Linguistics, pp. 1787–1796. Seattle, Washington, USA (2013). https://aclanthology.org/D13-1184

Chisholm, A., Hachey, B.: Entity disambiguation with web links. Trans. Assoc. Comput. Linguistics **3**, 145–156 (2015). https://doi.org/10.1162/tacl_a_00129. https://aclanthology.org/Q15-1011

Choi, E., Levy, O., Choi, Y., Zettlemoyer, L.: Ultra-fine entity typing. In: Proceedings of the 56th Annual Meeting of the Association for Computational Linguistics (Volume 1: Long Papers), Association for Computational Linguistics, pp. 87–96. Melbourne, Australia (2018). https://doi.org/10.18653/v1/P18-1009, https://aclanthology.org/P18-1009

Conneau, A., Khandelwal, K., Goyal, N., Chaudhary, V., Wenzek, G., Guzmán, F., Grave, E., Ott, M., Zettlemoyer, L., Stoyanov, V.: Unsupervised cross-lingual representation learning at scale. In: Proceedings of the 58th Annual Meeting of the Association for Computational Linguistics, pp. 8440–8451. Online (2020). https://doi.org/10.18653/v1/2020.acl-main.747, https://aclanthology.org/2020.acl-main.747

Cortes, C., Vapnik, V.: Support-vector networks. Machine Learn. **20**(3), 273–297 (1995)

Crichton, G., Pyysalo, S., Chiu, B., Korhonen, A.: A neural network multi-task learning approach to biomedical named entity recognition. BMC Bioinform. **18**(1), 368 (2017)

Cucerzan, S.: Large-scale named entity disambiguation based on Wikipedia data. In: Proceedings of the 2007 Joint Conference on Empirical Methods in Natural Language Processing and Computational Natural Language Learning (EMNLP-CoNLL), Association for Computational Linguistics, pp. 708–716. Prague, Czech Republic (2007). https://aclanthology.org/D07-1074

Darwish, K.: Named entity recognition using cross-lingual resources: arabic as an example. In: Proceedings of the 51st Annual Meeting of the Association for Computational Linguistics (Volume 1: Long Papers), Association for Computational Linguistics, pp. 1558–1567. Sofia, Bulgaria (2013). https://aclanthology.org/P13-1153

Das, D., Petrov, S.: Unsupervised part-of-speech tagging with bilingual graph-based projections. In: Proceedings of the 49th Annual Meeting of the Association for Computational Linguistics: Human Language Technologies, Association for Computational Linguistics, pp. 600–609. Portland, Oregon, USA (2011). https://aclanthology.org/P11-1061

De Cao, N., Wu, L., Popat, K., Artetxe, M., Goyal, N., Plekhanov, M., Zettlemoyer, L., Cancedda, N., Riedel, S., Petroni, F.: Multilingual autoregressive entity linking. Trans. Assoc. Comput. Linguistics **10**, 274–290 (2022). https://doi.org/10.1162/tacl_a_00460. https://aclanthology.org/2022.tacl-1.16

Delpeuch, A.: OpenTapioca: lightweight entity linking for Wikidata. arXiv preprint arXiv:1904.09131 (2019)

Demir, H., Özgür, A.: Improving named entity recognition for morphologically rich languages using word embeddings. In: 2014 13th International Conference on Machine Learning and Applications, pp. 117–122. IEEE (2014)

Dempster, A.P., Laird, N.M., Rubin, D.B.: Maximum likelihood from incomplete data via the EM algorithm. J. R. Stat. Soc., Series B (1977)

Derczynski, L., Bontcheva, K., Roberts, I.: Broad Twitter corpus: a diverse named entity recognition resource. In: Proceedings of COLING 2016, the 26th International Conference on Computational Linguistics: Technical Papers, The COLING 2016 Organizing Committee, pp. 1169–1179. Osaka, Japan (2016). https://aclanthology.org/C16-1111

Derczynski, L., Nichols, E., van Erp, M., Limsopatham, N.: Results of the WNUT2017 shared task on novel and emerging entity recognition. In: Proceedings of the 3rd Workshop on Noisy User-generated Text, Association for Computational Linguistics, pp. 140–147. Copenhagen, Denmark (2017). https://doi.org/10.18653/v1/W17-4418, https://aclanthology.org/W17-4418

Devlin, J., Chang, M.-W., Lee, K., Toutanova, K.: BERT: pre-training of deep bidirectional transformers for language understanding. In: Proceedings of the 2019 Conference of the North American Chapter of the Association for Computational Linguistics: Human Language Technologies, Volume 1 (Long and Short Papers), Association for Computational Linguistics, pp. 4171–4186. Minneapolis, Minnesota (2019). https://doi.org/10.18653/v1/N19-1423, https://aclanthology.org/N19-1423

Doddington, G., Mitchell, A., Przybocki, M., Ramshaw, L., Strassel, S., Weischedel, R.: The automatic content extraction (ACE) program—tasks, data, and evaluation. In: Proceedings of the Fourth International Conference on Language Resources and Evaluation (LREC'04), European Language Resources Association (ELRA). Lisbon, Portugal (2004). http://www.lrec-conf.org/proceedings/lrec2004/pdf/5.pdf

Dredze, M., McNamee, M., Rao, D., Gerber, A., Finin, T.: Entity disambiguation for knowledge base population. In: Proceedings of the 23rd International Conference on Computational Linguistics (Coling 2010), Coling 2010 Organizing Committee, pp. 277–285. Beijing, China (2010). https://aclanthology.org/C10-1032

Durrett, G., Klein, D.: A joint model for entity analysis: coreference, typing, and linking. Trans. Assoc. Comput. Linguistics **2**, 477–490 (2014). https://doi.org/10.1162/tacl_a_00197. https://aclanthology.org/Q14-1037

Dyer, C.: Notes on noise contrastive estimation and negative sampling. arXiv preprint arXiv:1410.8251 (2014)

Ehrmann, M., Turchi, M., Steinberger, R.: Building a multilingual named entity-annotated corpus using annotation projection. In: Proceedings of the International Conference Recent Advances in Natural Language Processing 2011, Association for Computational Linguistics, pp. 118–124. Hissar, Bulgaria (2011). https://aclanthology.org/R11-1017

Enghoff, J.V., Harrison, S., Agić, Ž.: Low-resource named entity recognition via multi-source projection: Not quite there yet? In: Proceedings of the 2018 EMNLP Workshop W-NUT: The 4th Workshop on Noisy User-generated Text, Association for Computational Linguistics, pp. 195–201. Brussels, Belgium (2018). https://doi.org/10.18653/v1/W18-6125, https://aclanthology.org/W18-6125

Färber, M., Ell, B., Menne, C., Rettinger, A.: A comparative survey of DBpedia, Freebase, OpenCyc, Wikidata, and YAGO. Semantic Web J. **1**(1), 1–5 (2015)

Faruqui, M., Dyer, C.: Improving vector space word representations using multilingual correlation. In: Proceedings of the 14th Conference of the European Chapter of the Association for Computational Linguistics, pp. 462–471. Gothenburg, Sweden (2014). https://doi.org/10.3115/v1/E14-1049, https://aclanthology.org/E14-1049

Ferragina, P., Scaiella, U.: TAGME: on-the-fly annotation of short text fragments (by Wikipedia entities). In: Proceedings of the ACM Conference on Information and Knowledge Management (CIKM) (2010)

Finch, A., Liu, L., Wang, X., Sumita, E.: Neural network transduction models in transliteration generation. In: Proceedings of the Fifth Named Entity Workshop, Association for Computational Linguistics, pp. 61–66. Beijing, China (2015). https://doi.org/10.18653/v1/W15-3909, https://aclanthology.org/W15-3909

Finkel, J.R., Grenager, T., Manning, C.: Incorporating non-local information into information extraction systems by Gibbs sampling. In: Proceedings of the 43rd Annual Meeting of the Association for Computational Linguistics (ACL'05), Association for Computational Linguistics, pp. 363–370. Ann Arbor, Michigan (2005). https://doi.org/10.3115/1219840.1219885, https://aclanthology.org/P05-1045

Francis-Landau, M., Durrett, G., Klein, D.: Capturing semantic similarity for entity linking with convolutional neural networks. In: Proceedings of the 2016 Conference of the North American Chapter of the Association for Computational Linguistics: Human Language Technologies, Association for Computational Linguistics, pp. 1256–1261. San Diego, California (2016). https://doi.org/10.18653/v1/N16-1150, https://aclanthology.org/N16-1150

Freitag, D.: Trained named entity recognition using distributional clusters. In: Proceedings of the 2004 Conference on Empirical Methods in Natural Language Processing, Association for Computational Linguistics, pp. 262–269. Barcelona, Spain (2004). https://aclanthology.org/W04-3234

Fu, X., Shi, W., Yu, X., Zhao, Z., Roth, D.: Design challenges in low-resource cross-lingual entity linking. In: Proceedings of the 2020 Conference on Empirical Methods in Natural Language Processing (EMNLP), Association for Computational Linguistics, pp. 6418–6432. Online (2020). https://doi.org/10.18653/v1/2020.emnlp-main.521, https://aclanthology.org/2020.emnlp-main.521

Ganchev, K., Gillenwater, J., Taskar, B.: Dependency grammar induction via bitext projection constraints. In: Proceedings of the Joint Conference of the 47th Annual Meeting of the ACL and the 4th International Joint Conference on Natural Language Processing of the AFNLP, Association for Computational Linguistics, pp. 369–377. Suntec, Singapore (2009). https://aclanthology.org/P09-1042

Ganea, O.-E., Hofmann, T.: Deep joint entity disambiguation with local neural attention. In: Proceedings of the 2017 Conference on Empirical Methods in Natural Language Processing, Association for Computational Linguistics, pp. 2619–2629. Copenhagen, Denmark (2017). https://doi.org/10.18653/v1/D17-1277, https://aclanthology.org/D17-1277

Ganea, O.-E., Ganea, M., Lucchi, A., Eickhoff, C., Hofmann, T.: Probabilistic bag-of-hyperlinks model for entity linking. In: Proceedings of the 25th International Conference on World Wide Web (WWW), pp. 927–938 (2016)

Gattani, A., Lamba, D.S., Garera, N., Tiwari, M., Chai, X., Das, S., Subramaniam, S., Rajaraman, A., Harinarayan, V., Doan, A.: Entity extraction, linking, classification, and tagging for social media: a Wikipedia-based approach. Very Large Data Base (VLDB) Endowment **6**(11), 1126–1137 (2013)

Getman, J., Ellis, J., Strassel, S., Song, Z., Tracey, J.: Laying the groundwork for knowledge base population: Nine years of linguistic resources for TAC KBP. In: Proceedings of the Eleventh International Conference on Language Resources and Evaluation (LREC 2018), European Language Resources Association (ELRA). Miyazaki, Japan (2018). https://aclanthology.org/L18-1245

Gillick, D., Kulkarni, S., Lansing, L., Presta, A., Baldridge, J., Ie, E., Garcia-Olano, D.: Learning dense representations for entity retrieval. In: Proceedings of the 23rd Conference on Computational Natural Language Learning (CoNLL), Association for Computational Linguistics, pp. 528–537. Hong Kong, China (2019). https://doi.org/10.18653/v1/K19-1049, https://aclanthology.org/K19-1049

Globerson, A., Lazic, N., Chakrabarti, S., Subramanya, A., Ringgaard, M., Pereira, F.: Collective entity resolution with multi-focal attention. In: Proceedings of the 54th Annual Meeting of the Association for Computational Linguistics (Volume 1: Long Papers), Association for Computational Linguistics, pp. 621–631. Berlin, Germany (2016). https://doi.org/10.18653/v1/P16-1059, https://aclanthology.org/P16-1059

Gottipati, S., Jiang, J.: Linking entities to a knowledge base with query expansion. In: Proceedings of the 2011 Conference on Empirical Methods in Natural Language Processing, Association for Computational Linguistics, pp. 804–813. Edinburgh, Scotland, UK (2011). https://aclanthology.org/D11-1074

Guha, R.V., Brickley, D., Macbeth, S.: Schema. org: evolution of structured data on the web. Commun. ACM **59**(2), 44–51 (2016)

Gui, T., Zou, Y., Zhang, Q., Peng, M., Fu, J., Wei, Z., Huang, X.: A lexicon-based graph neural network for Chinese NER. In: Proceedings of the 2019 Conference on Empirical Methods in Natural Language Processing and the 9th International Joint Conference on Natural Language Processing (EMNLP-IJCNLP), Association for Computational Linguistics, pp. 1040–1050. Hong Kong, China (2019). https://doi.org/10.18653/v1/D19-1096, https://aclanthology.org/D19-1096

Güngör, O., Uskudarli, S., Güngör, T.: Improving named entity recognition by jointly learning to disambiguate morphological tags. In: Proceedings of the 27th International Conference on Computational Linguistics, Association for Computational Linguistics, pp. 2082–2092. Santa Fe, New Mexico, USA (2018). https://aclanthology.org/C18-1177

Guo, S., Chang, M.-W., Kiciman, E.: To link or not to link? a study on end-to-end tweet entity linking. In: Proceedings of the 2013 Conference of the North American Chapter of the Association for Computational Linguistics: Human Language Technologies, Association for Computational Linguistics, pp. 1020–1030. Atlanta, Georgia (2013). https://aclanthology.org/N13-1122

Gupta, N., Singh, S., Roth, D.: Entity linking via joint encoding of types, descriptions, and context. In: Proceedings of the 2017 Conference on Empirical Methods in Natural Language Processing, Association for Computational Linguistics, pp. 2681–2690. Copenhagen, Denmark (2017). https://doi.org/10.18653/v1/D17-1284, https://aclanthology.org/D17-1284

Gutmann, M.U., Hyvärinen, A.: Noise-contrastive estimation of unnormalized statistical models, with applications to natural image statistics. J. Machine Learn. Res. (JMLR) **13**(Feb), 307–361 (2012)

Hachey, B., Radford, W., Nothman, J., Honnibal, M., Curran, J.R.: Evaluating entity linking with wikipedia. Artif. intell. **194**, 130–150 (2013)

Hajishirzi, H., Zilles, L., Weld, D.S., Zettlemoyer, L.: Joint coreference resolution and named-entity linking with multi-pass sieves. In: Proceedings of the 2013 Conference on Empirical Methods in Natural Language Processing, Association for Computational Linguistics, pp. 289–299. Seattle, Washington, USA (2013). https://aclanthology.org/D13-1029

Han, X., Sun, L.: A generative entity-mention model for linking entities with knowledge base. In: Proceedings of the 49th Annual Meeting of the Association for Computational Linguistics: Human Language Technologies, Association for Computational Linguistics, pp. 945–954. Portland, Oregon, USA (2011). https://aclanthology.org/P11-1095

Han, X., Sun, L.: An entity-topic model for entity linking. In: Proceedings of the 2012 Joint Conference on Empirical Methods in Natural Language Processing and Computational Natural Language Learning, Association for Computational Linguistics, pp. 105–115. Jeju Island, Korea (2012). https://aclanthology.org/D12-1010

Han, X., Sun, L., Zhao, J.: Collective entity linking in web text: a graph-based method. In: Proceedings of the ACM SIGIR Conference (SIGIR), pp. 765–774 (2011)

Hasan, K.S., Rahman, M.A.U., Ng, V.: Learning-based named entity recognition for morphologically-rich, resource-scarce languages. In: Proceedings of the 12th Conference of the European Chapter of the ACL (EACL 2009), Association for Computational Linguistics, pp. 354–362. Athens, Greece (2009). https://aclanthology.org/E09-1041

He, Z., Wang, H., Li, S.: The task 2 of CIPS-SIGHAN 2012 named entity recognition and disambiguation in Chinese bakeoff. In: Proceedings of the Second CIPS-SIGHAN Joint Conference on Chinese Language Processing, Association for Computational Linguistics, pp. 108–114. Tianjin, China (2012). https://aclanthology.org/W12-6321

He, Z., Liu, S., Li, M., Zhou, M., Zhang, L., Wang, H.: Learning entity representation for entity disambiguation. In: Proceedings of the 51st Annual Meeting of the Association for Computational Linguistics (Volume 2: Short Papers), Association for Computational Linguistics, pp. 30–34. Sofia, Bulgaria (2013). https://aclanthology.org/P13-2006

Herbrich, R., Graepel, T., Obermayer, K.: Large margin rank boundaries for ordinal regression (2000)

Hermann, K.M., Blunsom, P.: Multilingual distributed representations without word alignment. In: Proceedings of the International Conference on Learning Representations (2014)

Hochreiter, S., Schmidhuber, J.: Long short-term memory. Neural Comput. **9**(8), 1735–1780 (1997)

Hoffart, J., Yosef, M.A., Bordino, I., Fürstenau, H., Pinkal, M., Spaniol, M., Taneva, B., Thater, S., Weikum, G.: Robust disambiguation of named entities in text. In: Proceedings of the 2011 Conference on Empirical Methods in Natural Language Processing, Association for Computational Linguistics, pp. 782–792. Edinburgh, Scotland, UK (2011). https://aclanthology.org/D11-1072

Hoffart, J., Seufert, S., Nguyen, D.B., Theobald, M., Weikum, G.: KORE: keyphrase overlap relatedness for entity disambiguation. In: Proceedings of the ACM Conference on Information and Knowledge Management (CIKM), pp. 545–554. ACM (2012)

Hofstätter, S., Lin, S.-C., Yang, J.-H., Lin, J., Hanbury, A.: Efficiently teaching an effective dense retriever with balanced topic aware sampling. In: Proceedings of the 44th International ACM SIGIR Conference on Research and Development in Information Retrieval, pp. 113–122 (2021)

Hotelling, H.: Relations between two sets of variates. Biometrika **28**(3/4), 321–377 (1936). ISSN 00063444

Hovy, E., Marcus, M., Palmer, M., Ramshaw, L., Weischedel, R.: OntoNotes: the 90% solution. In: Proceedings of the Human Language Technology Conference of the NAACL, Companion Volume: Short Papers, Association for Computational Linguistics, pp. 57–60. New York City, USA (2006). https://aclanthology.org/N06-2015

Hu, J., Ruder, S., Siddhant, A., Neubig, G., Firat, O., Johnson, M.: Xtreme: a massively multilingual multi-task benchmark for evaluating cross-lingual generalisation. In: International Conference on Machine Learning, pp. 4411–4421. PMLR (2020)

Huang, Z., Xu, W., Yu, K.: Bidirectional LSTM-CRF models for sequence tagging. arXiv preprint arXiv:1508.01991 (2015)

Hwa, R., Resnik, P., Weinberg, A., Cabezas, C.I., Kolak, O.: Bootstrapping parsers via syntactic projection across parallel texts. Nat. Language Eng. (2005)

Ide, N., Véronis, J.: Introduction to the special issue on word sense disambiguation: the state of the art. Comput. Linguistics **24**(1), 1–40 (1998). https://aclanthology.org/J98-1001

Ihler, A.T., Fisher III, J.W., Willsky, A.S., Chickering, D.M.: Loopy belief propagation: convergence and effects of message errors. J. Machine Learn. Res. **6**(5) (2005)

Irvine, A., Callison-Burch, C., Klementiev, A.: Transliterating from all languages. In: Proceedings of the 9th Conference of the Association for Machine Translation in the Americas: Research Papers, Association for Machine Translation in the Americas. Denver, Colorado, USA (2010). https://aclanthology.org/2010.amta-papers.12

Islamaj Doğan, R., Lu, Z.: An improved corpus of disease mentions in PubMed citations. In: BioNLP: Proceedings of the 2012 Workshop on Biomedical Natural Language Processing, Association for Computational Linguistics, pp. 91–99. Montréal, Canada (2012). https://aclanthology.org/W12-2411

Jaleel, N.A., Larkey, L.S.: Statistical transliteration for English-Arabic cross language information retrieval. In: Proceedings of the ACM Conference on Information and Knowledge Management (CIKM) (2003)

Ji, H., Grishman, R.: Knowledge base population: successful approaches and challenges. In: Proceedings of the 49th Annual Meeting of the Association for Computational Linguistics: Human Language Technologies, Association for Computational Linguistics, pp. 1148–1158. Portland, Oregon, USA (2011). https://aclanthology.org/P11-1115

Ji, H., Grishman, R., Dang, H.T., Griffitt, K., Ellis, J.: Overview of the TAC 2010 knowledge base population track. In: Text Analysis Conference (TAC) (2010)

Ji, H., Grishman, R., Dang, H.T.: Overview of the TAC2011 knowledge base population track (2011)

Ji, H., Nothman, J., Hachey, B.: Overview of TAC-KBP2014 entity discovery and linking tasks. In: Text Analysis Conference (TAC) (2014)

Ji, H., Nothman, J., Hachey, B., Florian, R.: Overview of TAC-KBP2015 tri-lingual entity discovery and linking. In: Text Analysis Conference (TAC) (2015)

Ji, H., Nothman, J., Dang, H.T., Hub, S.I.: Overview of TAC-KBP2016 tri-lingual EDL and its impact on end-to-end cold-start KBP (2016)

Jiampojamarn, S., Bhargava, A., Dou, Q., Dwyer, K., Kondrak, G.: DirecTL: a language independent approach to transliteration. In: Proceedings of the 2009 Named Entities Workshop: Shared Task on Transliteration (NEWS 2009), Association for Computational Linguistics, pp. 28–31. Suntec, Singapore (2009). https://aclanthology.org/W09-3504

Jiampojamarn, S., Dwyer, K., Bergsma, S., Bhargava, A., Dou, Q., Kim, M.-Y., Kondrak, G.: Transliteration generation and mining with limited training resources. In: Proceedings of the 2010 Named Entities Workshop, Association for Computational Linguistics, pp. 39–47. Uppsala, Sweden (2010). https://aclanthology.org/W10-2405

Jin, G., Chen, X.: The fourth international Chinese language processing bakeoff: Chinese word segmentation, named entity recognition and Chinese POS tagging. In: Proceedings of the Sixth SIGHAN Workshop on Chinese Language Processing (2008). https://aclanthology.org/I08-4010

Joachims, T.: Unbiased evaluation of retrieval quality using clickthrough data. In: SIGIR Workshop on Mathematical/Formal Methods in Information Retrieval (2002)

Johnson, J., Douze, M., Jégou, H.: Billion-scale similarity search with GPUs. IEEE Trans. Big Data (2019)

Joko, H., Hasibi, F., Balog, K., de Vries, A.P.: Conversational entity linking: problem definition and datasets. In: Proceedings of the 44th International ACM SIGIR Conference on Research and Development in Information Retrieval, pp. 2390–2397 (2021)

Kamholz, D., Pool, J., Colowick, S.: PanLex: building a resource for panlingual lexical translation. In: Proceedings of the Ninth International Conference on Language Resources and Evaluation (LREC'14), European Language Resources Association (ELRA), pp. 3145–3150. Reykjavik, Iceland (2014). http://www.lrec-conf.org/proceedings/lrec2014/pdf/1029_Paper.pdf

Karimi, S., Scholer, F., Turpin, A.: Machine transliteration survey. ACM Comput. Surv. (2011). https://doi.org/10.1145/1922649.1922654

Karthikeyan, K., Wang, Z., Mayhew, S., Roth, D.: Cross-lingual ability of multilingual BERT: an empirical study. In: International Conference on Learning Representations (ICLR) (2019)

Khalid, M.A., Jijkoun, V., De Rijke, M.: The impact of named entity normalization on information retrieval for question answering. In: Proceedings of the European Conference on Information Retrieval (ECIR), pp. 705–710. Springer (2008)

Khashabi, D., Sammons, M., Zhou, B., Redman, T., Christodoulopoulos, C., Srikumar, V., Rizzolo, N., Ratinov, L., Luo, G., Do, Q., Tsai, C.-T., Roy, S., Mayhew, S., Feng, Z., Wieting, J., Yu, X., Song, Y., Gupta, S., Upadhyay, S., Arivazhagan, N., Ning, Q., Ling, S., Roth, D.: CogCompNLP: your Swiss army knife for NLP. In: Proceedings of the Eleventh International Conference on Language Resources and Evaluation (LREC 2018), European Language Resources Association (ELRA). Miyazaki, Japan (2018). https://aclanthology.org/L18-1086

Kim, S., Toutanova, K., Yu, H.: Multilingual named entity recognition using parallel data and metadata from Wikipedia. In: Proceedings of the 50th Annual Meeting of the Association for Computational Linguistics (Volume 1: Long Papers), Association for Computational Linguistics, pp. 694–702. Jeju Island, Korea (2012). https://aclanthology.org/P12-1073

Klementiev, A., Roth, D.: Named entity transliteration and discovery in multilingual corpora. In: Goutte, C., Cancedda, N., Dymetman, M., Foster, G. (eds.) Learning Machine Translation. MIT Press (2008). http://cogcomp.org/papers/KlementievRo08.pdf

Knight, K., Graehl, J.: Machine transliteration. In: 35th Annual Meeting of the Association for Computational Linguistics and 8th Conference of the European Chapter of the Association for Computational Linguistics, Association for Computational Linguistics, pp. 128–135. Madrid, Spain (1997). https://doi.org/10.3115/976909.979634, https://aclanthology.org/P97-1017

Koehn, P., Knowles, R.: Six challenges for neural machine translation. In: Proceedings of the First Workshop on Neural Machine Translation, Association for Computational Linguistics, pp. 28–39. Vancouver (2017). https://doi.org/10.18653/v1/W17-3204, https://aclanthology.org/W17-3204

Koehn, P., Hoang, H., Birch, A., Callison-Burch, C., Federico, M., Bertoldi, N., Cowan, B., Shen, W., Moran, C., Zens, R., Dyer, C., Bojar, O., Constantin, A., Herbst, E.: Moses: open source toolkit for statistical machine translation. In: Proceedings of the 45th Annual Meeting of the Association for Computational Linguistics Companion Volume Proceedings of the Demo and Poster Sessions, Association for Computational Linguistics, pp. 177–180. Prague, Czech Republic (2007). https://aclanthology.org/P07-2045

Kolitsas, N., Ganea, O.-E., Hofmann, T.: End-to-end neural entity linking. In: Proceedings of the 22nd Conference on Computational Natural Language Learning, Association for Computational Linguistics, pp. 519–529. Brussels, Belgium (2018). https://doi.org/10.18653/v1/K18-1050, https://aclanthology.org/K18-1050

Koo, T., Carreras, X., Collins, M.: Simple semi-supervised dependency parsing. In: Proceedings of ACL-08: HLT, Association for Computational Linguistics, pp. 595–603. Columbus, Ohio (2008). https://aclanthology.org/P08-1068

Kuhn, H.W.: The Hungarian method for the assignment problem. Naval Res. Logistics Q. **2**(1–2), 83–97 (1955)

Kulkarni, S., Singh, A., Ramakrishnan, G., Chakrabarti, S.: Collective annotation of Wikipedia entities in web text. In: Proceedings of the 15th ACM SIGKDD Conference on Knowledge Discovery and Data Mining (KDD), pp. 457–466. ACM (2009)

Lakoff, G., Johnson, M.: Metaphors We Live By. University of Chicago Press (1980)

Lakoff, G., Johnson, M.: Philosophy in the Flesh, vol. 4. Basic Books, New York (1999)

Lample, G., Conneau, A.: Cross-lingual language model pretraining. arXiv preprint arXiv:1901.07291 (2019)

Lample, G., Ballesteros, M., Subramanian, S., Kawakami, K., Dyer, C.: Neural architectures for named entity recognition. In: Proceedings of the 2016 Conference of the North American Chapter of the Association for Computational Linguistics: Human Language Technologies, Association for Computational Linguistics, pp. 260–270. San Diego, California (2016). https://doi.org/10.18653/v1/N16-1030. https://aclanthology.org/N16-1030

Lazic, N., Subramanya, A., Ringgaard, M., Pereira, F.: Plato: a selective context model for entity resolution. Trans. Assoc. Comput. Linguistics **3**, 503–515 (2015). https://doi.org/10.1162/tacl_a_00154. https://aclanthology.org/Q15-1036

Levy, O., Goldberg, Y.: Neural word embedding as implicit matrix factorization. In: Proceedings of the Conference on Advances in Neural Information Processing Systems (NIPS), pp. 2177–2185 (2014)

Lewis, M., Liu, Y., Goyal, N., Ghazvininejad, M., Mohamed, A., Levy, O., Stoyanov, V., Zettlemoyer, L.: BART: denoising sequence-to-sequence pre-training for natural language generation, translation, and comprehension. In: Proceedings of the 58th Annual Meeting of the Association for Computational Linguistics, pp. 7871–7880. Online (2020). https://doi.org/10.18653/v1/2020.acl-main.703, https://aclanthology.org/2020.acl-main.703

Li, B.Z., Min, S., Iyer, S., Mehdad, Y., Yih, W.-T.: Efficient one-pass end-to-end entity linking for questions. In: Proceedings of the 2020 Conference on Empirical Methods in Natural Language Processing (EMNLP), Association for Computational Linguistics, pp. 6433–6441. Online (2020a). https://doi.org/10.18653/v1/2020.emnlp-main.522, https://aclanthology.org/2020.emnlp-main.522

Li, H., Zhang, M., Su, J.: A joint source-channel model for machine transliteration. In: Proceedings of the 42nd Annual Meeting of the Association for Computational Linguistics (ACL-04), pp. 159–166. Barcelona, Spain (2004). https://doi.org/10.3115/1218955.1218976, https://aclanthology.org/P04-1021

Li, J., Sun, A., Han, J., Li, C.: A survey on deep learning for named entity recognition. IEEE Trans. Knowl. Data Eng. (2020b)

Li, Y., Wang, C., Han, F., Han, J., Roth, D., Yan, X.: Mining evidences for named entity disambiguation. In: Proceedings of the ACM SIGKDD Conference on Knowledge Discovery and Data Mining (KDD), pp. 1070–1078. ACM (2013)

Li, Z., Callison-Burch, C., Dyer, C., Khudanpur, S., Schwartz, L., Thornton, W., Weese, J., Zaidan, O.: Joshua: an open source toolkit for parsing-based machine translation. In: Proceedings of the Fourth Workshop on Statistical Machine Translation, Association for Computational Linguistics, pp. 135–139. Athens, Greece (2009). https://aclanthology.org/W09-0424

Lin, Y., Costello, C., Zhang, B., Lu, D., Ji, H., Mayfield, J., McNamee, P.: Platforms for non-speakers annotating names in any language. In: Proceedings of ACL 2018, System Demonstrations, Association for Computational Linguistics, pp. 1–6. Melbourne, Australia (2018). https://doi.org/10.18653/v1/P18-4001, https://aclanthology.org/P18-4001

Ling, X., Weld, D.S.: Fine-grained entity recognition. In: Proceedings of the National Conference on Artificial Intelligence (AAAI) (2012). http://aiweb.cs.washington.edu/ai/pubs/ling-aaai12.pdf

Ling, X., Singh, S., Weld, D.S.: Design challenges for entity linking. Trans. Assoc. Comput. Linguistics **3**, 315–328 (2015). https://doi.org/10.1162/tacl_a_00141. https://aclanthology.org/Q15-1023

Liu, X., Li, Y., Wu, H., Zhou, M., Wei, F., Lu, Y.: Entity linking for tweets. In: Proceedings of the 51st Annual Meeting of the Association for Computational Linguistics (Volume 1: Long Papers), Association for Computational Linguistics, pp. 1304–1311. Sofia, Bulgaria (2013). https://aclanthology.org/P13-1128

Liu, Y., Ott, M., Goyal, N., Du, J., Joshi, M., Chen, D., Levy, O., Lewis, M., Zettlemoyer, L., Stoyanov, V.: RoBERTa: a robustly optimized BERT pretraining approach. arXiv preprint arXiv:1907.11692, abs/1907.11692 (2019)

Liu, Y., Gu, J., Goyal, N., Li, X., Edunov, S., Ghazvininejad, M., Lewis, M., Zettlemoyer, L.: Multilingual denoising pre-training for neural machine translation. Trans. Assoc. Comput. Linguistics **8**, 726–742 (2020). https://doi.org/10.1162/tacl_a_00343. https://aclanthology.org/2020.tacl-1.47

Logeswaran, L., Chang, M.-W., Lee, K., Toutanova, K., Devlin, J., Lee, H.: Zero-shot entity linking by reading entity descriptions. In: Proceedings of the 57th Annual Meeting of the Association for Computational Linguistics, Association for Computational Linguistics, pp. 3449–3460. Florence, Italy (2019). https://doi.org/10.18653/v1/P19-1335, https://aclanthology.org/P19-1335

Lopez, A., Post, M.: Beyond bitext: five open problems in machine translation. In: Proceedings of the EMNLP Workshop on Twenty Years of Bitext (2013)

Lu, A., Wang, W., Bansal, M., Gimpel, K., Livescu, K.: Deep multilingual correlation for improved word embeddings. In: Proceedings of the 2015 Conference of the North American Chapter of the Association for Computational Linguistics: Human Language Technologies, Association for Computational Linguistics, pp. 250–256. Denver, Colorado (2015). https://doi.org/10.3115/v1/N15-1028, https://aclanthology.org/N15-1028

Luo, G., Huang, X., Lin, C.-Y., Nie, Z.: Joint entity recognition and disambiguation. In: Proceedings of the 2015 Conference on Empirical Methods in Natural Language Processing, Association for Computational Linguistics, pp. 879–888. Lisbon, Portugal (2015). https://doi.org/10.18653/v1/D15-1104, https://aclanthology.org/D15-1104

Luo, X.: On coreference resolution performance metrics. In: Proceedings of Human Language Technology Conference and Conference on Empirical Methods in Natural Language Processing, Association for Computational Linguistics, pp. 25–32. Vancouver, British Columbia, Canada (2005). https://aclanthology.org/H05-1004

Luong, T., Pham, H., Manning, C.D.: Bilingual word representations with monolingual quality in mind. In: Proceedings of the 1st Workshop on Vector Space Modeling for Natural Language Processing, Association for Computational Linguistics, pp. 151–159. Denver, Colorado (2015). https://doi.org/10.3115/v1/W15-1521, https://aclanthology.org/W15-1521

Ma, X., Hovy, E.: End-to-end sequence labeling via bi-directional LSTM-CNNs-CRF. In: Proceedings of the 54th Annual Meeting of the Association for Computational Linguistics (Volume 1: Long Papers), Association for Computational Linguistics, pp. 1064–1074. Berlin, Germany (2016). https://doi.org/10.18653/v1/P16-1101, https://aclanthology.org/P16-1101

Manning, C., Surdeanu, M., Bauer, J., Finkel, J., Bethard, S., McClosky, D.: The Stanford CoreNLP natural language processing toolkit. In: Proceedings of 52nd Annual Meeting of the Association for Computational Linguistics: System Demonstrations, Association for Computational Linguistics, pp. 55–60. Baltimore, Maryland (2014). https://doi.org/10.3115/v1/P14-5010, https://aclanthology.org/P14-5010

Mao, X., Dong, Y., He, S., Bao, S., Wang, H.: Chinese word segmentation and named entity recognition based on conditional random fields. In: Proceedings of the Sixth SIGHAN Workshop on Chinese Language Processing (2008). https://aclanthology.org/I08-4013

Mayhew, S., Roth, D.: TALEN: tool for annotation of low-resource ENtities. In: Proceedings of ACL 2018, System Demonstrations, Association for Computational Linguistics, pp. 80–86. Melbourne, Australia (2018). https://doi.org/10.18653/v1/P18-4014, https://aclanthology.org/P18-4014

Mayhew, S., Tsai, C.-T., Roth, D.: Cheap translation for cross-lingual named entity recognition. In: Proceedings of the 2017 Conference on Empirical Methods in Natural Language Processing, Association for Computational Linguistics, pp. 2536–2545. Copenhagen, Denmark (2017). https://doi.org/10.18653/v1/D17-1269, https://aclanthology.org/D17-1269

McCallum, A., Li, W.: Early results for named entity recognition with conditional random fields, feature induction and web-enhanced lexicons. In: Proceedings of the Seventh Conference on Natural Language Learning at HLT-NAACL 2003, pp. 188–191 (2003). https://aclanthology.org/W03-0430

McDonald, R., Petrov, S., Hall, K.: Multi-source transfer of delexicalized dependency parsers. In: Proceedings of the 2011 Conference on Empirical Methods in Natural Language Processing, Association for Computational Linguistics, pp. 62–72. Edinburgh, Scotland, UK (2011). https://aclanthology.org/D11-1006

McNamee, P., Dang, H.T.: Overview of the TAC 2009 knowledge base population track. In: Text Analysis Conference (TAC), vol. 17, pp. 111–113 (2009)

McNamee, P., Mayfield, J.: Entity extraction without language-specific resources. In: COLING-02: The 6th Conference on Natural Language Learning 2002 (CoNLL-2002) (2002). https://aclanthology.org/W02-2020

McNamee, P., Mayfield, J., Lawrie, D., Oard, D., Doermann, D.: Cross-language entity linking. In: Proceedings of 5th International Joint Conference on Natural Language Processing, Asian Federation of Natural Language Processing, pp. 255–263. Chiang Mai, Thailand (2011). https://aclanthology.org/I11-1029

Mendes, P.N., Jakob, M., García-Silva, A., Bizer, C.: DBpedia spotlight: shedding light on the web of documents. In: Proceedings of the 7th International Conference on Semantic Systems, pp. 1–8. ACM (2011)

Mihalcea, R.: Using Wikipedia for automatic word sense disambiguation. In: Human Language Technologies 2007: The Conference of the North American Chapter of the Association for Computational Linguistics; Proceedings of the Main Conference, Association for Computational Linguistics, pp. 196–203. Rochester, New York (2007). https://aclanthology.org/N07-1025

Mihalcea, R., Csomai, A.: Wikify!: linking documents to encyclopedic knowledge. In: Proceedings of the ACM Conference on Information and Knowledge Management (CIKM) (2007)

Mikolov, T., Chen, K., Corrado, G., Dean, J.: Efficient estimation of word representations in vector space. In: Proceedings of the International Conference on Learning Representations (2013a)

Mikolov, T., Le, Q.V., Sutskever, I.: Exploiting similarities among languages for machine translation. arXiv preprint arXiv:1309.4168 (2013b)

Mikolov, T., Sutskever, I., Chen, K., Corrado, G.S., Dean, J.: Distributed representations of words and phrases and their compositionality. In: Proceedings of the Conference on Advances in Neural Information Processing Systems (NIPS) (2013c)

Miller, G., Beckwith, R., Fellbaum, C., Gross, D., Miller, K.: Wordnet: an on-line lexical database. Int. J. Lexicography (1990)

Miller, S., Guinness, J., Zamanian, A.: Name tagging with word clusters and discriminative training. In: Proceedings of the Human Language Technology Conference of the North American Chapter of the Association for Computational Linguistics: HLT-NAACL 2004, Association for Computational Linguistics, pp. 337–342. Boston, Massachusetts, USA (2004). https://aclanthology.org/N04-1043

Milne, D., Witten, I.H.: Learning to link with Wikipedia. In: Proceedings of the ACM Conference on Information and Knowledge Management (CIKM) (2008)

Moro, A., Raganato, A., Navigli, R.: Entity linking meets word sense disambiguation: a unified approach. Trans. Assoc. Comput. Linguistics **2**, 231–244 (2014). https://doi.org/10.1162/tacl_a_00179. https://aclanthology.org/Q14-1019

Mulang, I.O., Singh, K., Vyas, A., Shekarpour, S., Sakor, A., Vidal, M.E., Auer, S., Lehmann, J.: Context-aware entity linking with attentive neural networks on Wikidata knowledge graph (2019)

Murphy, K., Weiss, Y., Jordan, M.I.: Loopy belief propagation for approximate inference: an empirical study. arXiv preprint arXiv:1301.6725 (2013)

Nadeau, D., Sekine, S.: A survey of named entity recognition and classification. Lingvisticae Investigationes **30**(1), 3–26 (2007)

Nguyen, D.B., Theobald, M., Weikum, G.: J-NERD: joint named entity recognition and disambiguation with rich linguistic features. Trans. Assoc. Comput. Linguistics **4**, 215–229 (2016). https://doi.org/10.1162/tacl_a_00094. https://aclanthology.org/Q16-1016

Nie, Y., Tian, Y., Wan, X., Song, Y., Dai, B.: Named entity recognition for social media texts with semantic augmentation. In: Proceedings of the 2020 Conference on Empirical Methods in Natural Language Processing (EMNLP), Association for Computational Linguistics, pp. 1383–1391. Online(2020). https://doi.org/10.18653/v1/2020.emnlp-main.107, https://aclanthology.org/2020.emnlp-main.107

Nothman, J., Curran, J.R., Murphy, T.: Transforming Wikipedia into named entity training data. In: Proceedings of the Australasian Language Technology Association Workshop 2008, pp. 124–132. Hobart, Australia (2008). https://aclanthology.org/U08-1016

Nothman, J., Murphy, T., Curran, J.R.: Analysing Wikipedia and gold-standard corpora for NER training. In: Proceedings of the 12th Conference of the European Chapter of the ACL (EACL 2009), Association for Computational Linguistics, pp. 612–620. Athens, Greece (2009). https://aclanthology.org/E09-1070

Nothman, J., Ringland, N., Radford, W., Murphy, T., Curran, J.R.: Learning multilingual named entity recognition from Wikipedia. Artif. Intell. **194**, 151–175 (2013)

Och, F.J., Ney, H.: A systematic comparison of various statistical alignment models. Comput. Linguistics **29**(1), 19–51 (2003). https://doi.org/10.1162/089120103321337421. https://aclanthology.org/J03-1002

Pan, X., Cassidy, T., Hermjakob, U., Ji, H., Knight, K.: Unsupervised entity linking with abstract meaning representation. In: Proceedings of the 2015 Conference of the North American Chapter of the Association for Computational Linguistics: Human Language Technologies, Association for Computational Linguistics, pp. 1130–1139. Denver, Colorado (2015). https://doi.org/10.3115/v1/N15-1119, https://aclanthology.org/N15-1119

Pan, X., Zhang, B., May, J., Nothman, J., Knight, K., Ji, H.: Cross-lingual name tagging and linking for 282 languages. In: Proceedings of the 55th Annual Meeting of the Association for Computational Linguistics (Volume 1: Long Papers), Association for Computational Linguistics, pp. 1946–1958. Vancouver, Canada (2017). https://doi.org/10.18653/v1/P17-1178, https://aclanthology.org/P17-1178

Pasternack, J., Roth, D.: Learning better transliterations. In: Proceedings of the ACM Conference on Information and Knowledge Management (CIKM) (2009). http://cogcomp.org/papers/PasternackRo09a.pdf

Pellissier Tanon, T., Weikum, G., Suchanek, F.: YAGO 4: a reason-able knowledge base. In: The Semantic Web: 17th International Conference, pp. 583–596. Springer (2020)

Peng, N., Dredze, M.: Named entity recognition for Chinese social media with jointly trained embeddings. In: Proceedings of the 2015 Conference on Empirical Methods in Natural Language Processing, Association for Computational Linguistics, pp. 548–554. Lisbon, Portugal (2015). https://doi.org/10.18653/v1/D15-1064, https://aclanthology.org/D15-1064

Peng, N., Dredze. M.: Improving named entity recognition for Chinese social media with word segmentation representation learning. In: Proceedings of the 54th Annual Meeting of the Association for Computational Linguistics (Volume 2: Short Papers), Association for Computational Linguistics, pp. 149–155. Berlin, Germany (2016). https://doi.org/10.18653/v1/P16-2025, https://aclanthology.org/P16-2025

Pennington, J., Socher, R., Manning, C.: GloVe: global vectors for word representation. In: Proceedings of the 2014 Conference on Empirical Methods in Natural Language Processing (EMNLP), Association for Computational Linguistics, pp. 1532–1543. Doha, Qatar (2014). https://doi.org/10.3115/v1/D14-1162, https://aclanthology.org/D14-1162

Pershina, M., He, Y., Grishman, R.: Personalized page rank for named entity disambiguation. In: Proceedings of the 2015 Conference of the North American Chapter of the Association for Computational Linguistics: Human Language Technologies, Association for Computational Linguistics, pp. 238–243. Denver, Colorado (2015). https://doi.org/10.3115/v1/N15-1026, https://aclanthology.org/N15-1026

Peters, M.E., Neumann, M., Iyyer, M., Gardner, M., Clark, C., Lee, K., Zettlemoyer, L.: Deep contextualized word representations. In: Proceedings of the 2018 Conference of the North American Chapter of the Association for Computational Linguistics: Human Language Technologies, Volume 1 (Long Papers), Association for Computational Linguistics, pp. 2227–2237. New Orleans, Louisiana (2018). https://doi.org/10.18653/v1/N18-1202, https://aclanthology.org/N18-1202

Pilz, A., Paaß, G.: From names to entities using thematic context distance. In: Proceedings of the ACM Conference on Information and Knowledge Management (CIKM), pp. 857–866 (2011)

Pingali, P., Ganesh, S., Yella, S., Varma, V.: Statistical transliteration for cross language information retrieval using HMM alignment model and CRF. In: Proceedings of the 2nd workshop on Cross Lingual Information Access (CLIA) Addressing the Information Need of Multilingual Societies (2008). https://aclanthology.org/I08-6006

Pires, T., Schlinger, E., Garrette, D.: How multilingual is multilingual BERT? In: Proceedings of the 57th Annual Meeting of the Association for Computational Linguistics (ACL), Association for Computational Linguistics, pp. 4996–5001. Florence, Italy (2019). https://doi.org/10.18653/v1/P19-1493, https://www.aclweb.org/anthology/P19-1493

Ponzetto S.P., Strube, M.: Exploiting semantic role labeling, WordNet and Wikipedia for coreference resolution. In: Proceedings of the Human Language Technology Conference of the NAACL, Main Conference, Association for Computational Linguistics, pp. 192–199. New York City, USA (2006). https://aclanthology.org/N06-1025

Raffel, C., Shazeer, N., Roberts, A., Lee, K., Narang, S., Matena, M., Zhou, Y., Li, W., Liu, P.J., et al.: Exploring the limits of transfer learning with a unified text-to-text transformer. The J. Machine Learn. Res. **21**(140), 1–67 (2020)

Rahman, A., Ng, V.: Coreference resolution with world knowledge. In: Proceedings of the 49th Annual Meeting of the Association for Computational Linguistics: Human Language Technologies, Association for Computational Linguistics, pp. 814–824. Portland, Oregon, USA (2011). https://aclanthology.org/P11-1082

Ratinov, L., Roth, D.: Design challenges and misconceptions in named entity recognition. In: Proceedings of the Thirteenth Conference on Computational Natural Language Learning (CoNLL-2009), Association for Computational Linguistics, pp. 147–155. Boulder, Colorado (2009). https://aclanthology.org/W09-1119

Ratinov, L., Roth, D.: Learning-based multi-sieve co-reference resolution with knowledge. In: Proceedings of the 2012 Joint Conference on Empirical Methods in Natural Language Processing and Computational Natural Language Learning, Association for Computational Linguistics, pp. 1234–1244. Jeju Island, Korea (2012). https://aclanthology.org/D12-1113

Ratinov, L., Roth, D., Downey, D., Anderson, M.: Local and global algorithms for disambiguation to Wikipedia. In: Proceedings of the 49th Annual Meeting of the Association for Computational Linguistics: Human Language Technologies, Association for Computational Linguistics, pp. 1375–1384. Portland, Oregon, USA (2011). https://aclanthology.org/P11-1138

Ravi, S., Knight, K.: Learning phoneme mappings for transliteration without parallel data. In: Proceedings of Human Language Technologies: The 2009 Annual Conference of the North American

Chapter of the Association for Computational Linguistics, Association for Computational Linguistics, pp. 37–45. Boulder, Colorado (2009). https://aclanthology.org/N09-1005

Rebele, T., Suchanek, F., Hoffart, J., Biega, J., Kuzey, E., Weikum, G.: YAGO: a multilingual knowledge base from Wikipedia, Wordnet, and Geonames. In: International Semantic Web Conference, pp. 177–185. Springer (2016)

Reddy, S., Waxmonsky, S.: Substring-based transliteration with conditional random fields. In: Proceedings of the 2009 Named Entities Workshop: Shared Task on Transliteration (NEWS 2009), Association for Computational Linguistics, pp. 92–95. Suntec, Singapore (2009). https://aclanthology.org/W09-3520

Reimers, N., Gurevych, I.: Reporting score distributions makes a difference: performance study of LSTM-networks for sequence tagging. In: Proceedings of the 2017 Conference on Empirical Methods in Natural Language Processing, Association for Computational Linguistics, pp. 338–348. Copenhagen, Denmark (2017). https://doi.org/10.18653/v1/D17-1035, https://aclanthology.org/D17-1035

Reimers, N., Gurevych, I.: Sentence-BERT: sentence embeddings using Siamese BERT-networks. In: Proceedings of the 2019 Conference on Empirical Methods in Natural Language Processing and the 9th International Joint Conference on Natural Language Processing (EMNLP-IJCNLP), Association for Computational Linguistics, pp. 3982–3992. Hong Kong, China (2019). https://doi.org/10.18653/v1/D19-1410, https://aclanthology.org/D19-1410

Rijhwani, S., Xie, J., Neubig, G., Carbonell, J.: Zero-shot neural transfer for cross-lingual entity linking. In: Proceedings of the AAAI Conference on Artificial Intelligence, vol. 33, pp. 6924–6931 (2019)

Rijhwani, S., Zhou, S., Neubig, G., Carbonell, J.: Soft gazetteers for low-resource named entity recognition. In: Proceedings of the 58th Annual Meeting of the Association for Computational Linguistics, pp. 8118–8123. Online (2020). https://doi.org/10.18653/v1/2020.acl-main.722, https://aclanthology.org/2020.acl-main.722

Robertson, S.E., Walker, S.: Some simple effective approximations to the 2-poisson model for probabilistic weighted retrieval. In: Proceedings of the Seventeenth Annual International ACM-SIGIR Conference on Research and Development in Information Retrieval, pp. 232–241. Springer (1994)

Rosenbaltt, F.: The perceptron—a perciving and recognizing automation. Cornell Aeronautical Lab. (1957)

Roth, D., Yih, W.-T.: A linear programming formulation for global inference in natural language tasks. In: Proceedings of the Eighth Conference on Computational Natural Language Learning (CoNLL-2004) at HLT-NAACL 2004, Association for Computational Linguistics, pp. 1–8. Boston, Massachusetts, USA (2004). https://aclanthology.org/W04-2401

Sahlgren, M.: The word-space model: using distributional analysis to represent syntagmatic and paradigmatic relations between words in high-dimensional vector spaces. Ph.D. thesis, Stockholm University (2006)

Sakor, A., Singh, K., Vidal, M.: FALCON: an entity and relation linking framework over DBpedia. In: Proceedings of the ISWC 2019 Satellite Tracks, pp. 265–268 (2019)

Sarawagi, S., Cohen, W.W.: Semi-Markov conditional random fields for information extraction. In: Proceedings of the Conference on Advances in Neural Information Processing Systems (NIPS) (2004)

Şeker, G.A., Eryiğit, G.: Initial explorations on using CRFs for Turkish named entity recognition. In: Proceedings of COLING 2012, The COLING 2012 Organizing Committee, pp. 2459–2474. Mumbai, India (2012). https://aclanthology.org/C12-1150

Shen, W., Wang, J., Luo, P., Wang, M.: LINDEN: linking named entities with knowledge base via semantic knowledge. In: Proceedings of the 21st international conference on World Wide Web (WWW), pp. 449–458. ACM (2012)

Shen, W., Wang, J., Han, J.: Entity linking with a knowledge base: issues, techniques, and solutions. IEEE Trans. Knowl. Data Eng. **27**(2), 443–460 (2015)

Sil, A., Florian, R.: One for all: Towards language independent named entity linking. In: Proceedings of the 54th Annual Meeting of the Association for Computational Linguistics (Volume 1: Long Papers), Association for Computational Linguistics, pp. 2255–2264. Berlin, Germany (2016). https://doi.org/10.18653/v1/P16-1213, https://aclanthology.org/P16-1213

Sil, A., Yates, A.: Re-ranking for joint named-entity recognition and linking. In: Proceedings of the 22nd ACM International Conference on Information & Knowledge Management (CIKM), pp. 2369–2374. ACM (2013)

Sil, A., Kundu, G., Florian, R., Hamza, W.: Neural cross-lingual entity linking. In: Proceedings of the Conference on Artificial Intelligence (AAAI) (2018)

Smith, S.L., Turban, D.H., Hamblin, S., Hammerla, N.Y.: Offline bilingual word vectors, orthogonal transformations and the inverted softmax. In: Proceedings of the International Conference on Learning Representations (ICLR) (2017)

Søgaard, A., Vulić, I., Ruder, S., Faruqui, M.: Cross-lingual word embeddings. Synthesis Lectures on Human Language Technol. **12**(2), 1–132 (2019)

Spitkovsky, V.I., Chang, A.X.: A cross-lingual dictionary for English Wikipedia concepts. In: Proceedings of the Eighth International Conference on Language Resources and Evaluation (LREC'12), European Language Resources Association (ELRA), pp. 3168–3175. Istanbul, Turkey (2012). http://www.lrec-conf.org/proceedings/lrec2012/pdf/266_Paper.pdf

Sproat, R., Tao, T., Zhai, C.: Named entity transliteration with comparable corpora. In: Proceedings of the 21st International Conference on Computational Linguistics and 44th Annual Meeting of the Association for Computational Linguistics, Association for Computational Linguistics, pp. 73–80. Sydney, Australia (2006). https://doi.org/10.3115/1220175.1220185, https://aclanthology.org/P06-1010

Strubell, E., Verga, P., Belanger, D., McCallum, A.: Fast and accurate entity recognition with iterated dilated convolutions. In: Proceedings of the 2017 Conference on Empirical Methods in Natural Language Processing, Association for Computational Linguistics, pp. 2670–2680. Copenhagen, Denmark (2017). https://doi.org/10.18653/v1/D17-1283. https://aclanthology.org/D17-1283

Suchanek, F.M., Kasneci, G., Weikum, G.: YAGO: a core of semantic knowledge. In: Proceedings of the International World Wide Web Conference (WWW) (2007)

Sun, H., Ma, H., Yih, W.-T., Tsai, C.-T., Liu, J., Chang, M.-W.: Open domain question answering via semantic enrichment. In: Proceedings of the International World Wide Web Conference (WWW), International World Wide Web Conferences Steering Committee, pp. 1045–1055 (2015)

Sutton, C., McCallum, A.: An introduction to conditional random fields for relational learning. Introduction Stat. Relational Learn. **2**, 93–128 (2006)

Täckström, O., McDonald, R., Uszkoreit, J.: Cross-lingual word clusters for direct transfer of linguistic structure. In: Proceedings of the 2012 Conference of the North American Chapter of the Association for Computational Linguistics: Human Language Technologies, Association for Computational Linguistics, pp. 477–487. Montréal, Canada (2012). https://aclanthology.org/N12-1052

Tjong Kim Sang, E.F.: Introduction to the CoNLL-2002 shared task: language-independent named entity recognition. In: COLING-02: The 6th Conference on Natural Language Learning 2002 (CoNLL-2002) (2002). https://aclanthology.org/W02-2024

Tjong Kim Sang, E.F., De Meulder, F.: Introduction to the CoNLL-2003 shared task: language-independent named entity recognition. In: Proceedings of the Seventh Conference on Natural Language Learning at HLT-NAACL 2003, pp. 142–147 (2003). https://aclanthology.org/W03-0419

Tracey, J., Strassel, S.: Basic language resources for 31 languages (plus English): The LORELEI representative and incident language packs. In: Proceedings of the 1st Joint Workshop on Spoken

Language Technologies for Under-resourced languages (SLTU) and Collaboration and Computing for Under-Resourced Languages (CCURL), European Language Resources association, pp. 277–284. Marseille, France (2020). ISBN 979-10-95546-35-1. https://aclanthology.org/2020.sltu-1.39

Tracey, J., Strassel, S., Bies, A., Song, Z., Arrigo, M., Griffitt, K., Delgado, D., Graff, D., Kulick, S., Mott, J., Kuster, N.: Corpus building for low resource languages in the DARPA LORELEI program. In: Proceedings of the 2nd Workshop on Technologies for MT of Low Resource Languages, European Association for Machine Translation, pp. 48–55. Dublin, Ireland (2019). https://aclanthology.org/W19-6808

Tratz, S., Hovy, E.: A fast, accurate, non-projective, semantically-enriched parser. In: Proceedings of the 2011 Conference on Empirical Methods in Natural Language Processing, Association for Computational Linguistics, pp. 1257–1268. Edinburgh, Scotland, UK (2011). https://aclanthology.org/D11-1116

Tsai, C.-T., Roth, D.: Concept grounding to multiple knowledge bases via indirect supervision. Trans. Assoc. Comput. Linguistics **4**, 141–154 (2016). https://doi.org/10.1162/tacl_a_00089. https://aclanthology.org/Q16-1011

Tsai, C.-T., Roth, D.: Cross-lingual wikification using multilingual embeddings. In: Proceedings of the 2016 Conference of the North American Chapter of the Association for Computational Linguistics: Human Language Technologies, Association for Computational Linguistics, pp. 589–598. San Diego, California (2016b). https://doi.org/10.18653/v1/N16-1072, https://aclanthology.org/N16-1072

Tsai, C.-T., Roth, D.: Learning better name translation for cross-lingual Wikification. In: Proceedings of the Conference on Artificial Intelligence (AAAI) (2018). http://cogcomp.org/papers/TsaiRo18.pdf

Tsai, C.-T., Mayhew, S., Roth, D.: Cross-lingual named entity recognition via wikification. In: Proceedings of the 20th SIGNLL Conference on Computational Natural Language Learning, Association for Computational Linguistics, pp. 219–228. Berlin, Germany (2016). https://doi.org/10.18653/v1/K16-1022, https://aclanthology.org/K16-1022

Tür, G.: A statistical information extraction system for Turkish. Ph.D. thesis, Bilkent University (2000)

Turian, J., Ratinov, L.-A., Bengio, Y.: Word representations: a simple and general method for semi-supervised learning. In: Proceedings of the 48th Annual Meeting of the Association for Computational Linguistics, pp. 384–394. Uppsala, Sweden (2010). https://aclanthology.org/P10-1040

Upadhyay, S., Gupta, N., Roth, D.: Joint multilingual supervision for cross-lingual entity linking. In: Proceedings of the 2018 Conference on Empirical Methods in Natural Language Processing, Association for Computational Linguistics, pp. 2486–2495. Brussels, Belgium (2018). https://doi.org/10.18653/v1/D18-1270, https://aclanthology.org/D18-1270

Vilain, M., Burger, J., Aberdeen, J., Connolly, D., Hirschman, L.: A model-theoretic coreference scoring scheme. In: Sixth Message Understanding Conference (MUC-6): Proceedings of a Conference Held in Columbia, Maryland, November 6-8, 1995 (1995). https://aclanthology.org/M95-1005

Virga, P., Khudanpur, S.: Transliteration of proper names in cross-lingual information retrieval. In: Proceedings of the ACL 2003 Workshop on Multilingual and Mixed-language Named Entity Recognition, Association for Computational Linguistics, pp. 57–64. Sapporo, Japan (2003). https://doi.org/10.3115/1119384.1119392, https://aclanthology.org/W03-1508

Vrandečić, D.: Wikidata: a new platform for collaborative data collection. In: Proceedings of the International World Wide Web Conference (WWW), pp. 1063–1064. ACM (2012)

Wang, H., Zheng, J.G., Ma, X., Fox, P., Ji, H.: Language and domain independent entity linking with quantified collective validation. In: Proceedings of the 2015 Conference on Empirical Methods in

Natural Language Processing, Association for Computational Linguistics, pp. 695–704. Lisbon, Portugal (2015). https://doi.org/10.18653/v1/D15-1081, https://aclanthology.org/D15-1081

Wang, M., Manning, C.D.: Cross-lingual projected expectation regularization for weakly supervised learning. Trans. Assoc. Comput. Linguistics **2**, 55–66 (2014). https://doi.org/10.1162/tacl_a_00165. https://aclanthology.org/Q14-1005

Wang, Z., Zhang, J., Feng, J., Chen, Z.: Knowledge graph and text jointly embedding. In: Proceedings of the 2014 Conference on Empirical Methods in Natural Language Processing (EMNLP), Association for Computational Linguistics, pp. 1591–1601. Doha, Qatar (2014). https://doi.org/10.3115/v1/D14-1167, https://aclanthology.org/D14-1167

Weaver, W.: Translation. In: Proceedings of the Conference on Mechanical Translation, Massachusetts Institute of Technology (1952). https://aclanthology.org/1952.earlymt-1.1

Weischedel, R., Palmer, M., Marcus, M., Hovy, E., Pradhan, S., Ramshaw, L., Xue, N., Taylor, A., Kaufman, J., Franchini, M., et al.: Ontonotes release 5.0. Linguistic Data Consortium, Philadelphia, PA, vol. 23 (2013)

Wendemuth, A.: Learning the unlearnable. J. Phys. A: Math. General **28**(18), 5423 (1995)

Wu, L., Petroni, F., Josifoski, M., Riedel, S., Zettlemoyer, L.: Scalable zero-shot entity linking with dense entity retrieval. In: Proceedings of the 2020 Conference on Empirical Methods in Natural Language Processing (EMNLP), Association for Computational Linguistics, pp. 6397–6407. Online (2020). https://doi.org/10.18653/v1/2020.emnlp-main.519, https://aclanthology.org/2020.emnlp-main.519

Wu, S., Dredze, M.: Beto, bentz, becas: the surprising cross-lingual effectiveness of BERT. In: Proceedings of the 2019 Conference on Empirical Methods in Natural Language Processing and the 9th International Joint Conference on Natural Language Processing (EMNLP-IJCNLP), Association for Computational Linguistics, pp. 833–844. Hong Kong, China (2019). https://doi.org/10.18653/v1/D19-1077, https://www.aclweb.org/anthology/D19-1077

Xue, L., Constant, N., Roberts, A., Kale, M., Al-Rfou, R., Siddhant, A., Barua, A., Raffel, C.: mT5: a massively multilingual pre-trained text-to-text transformer. In: Proceedings of the 2021 Conference of the North American Chapter of the Association for Computational Linguistics: Human Language Technologies, Association for Computational Linguistics, pp. 483–498. Online (2021). https://doi.org/10.18653/v1/2021.naacl-main.41, https://aclanthology.org/2021.naacl-main.41

Yadav, V., Bethard, S.: A survey on recent advances in named entity recognition from deep learning models. In: Proceedings of the 27th International Conference on Computational Linguistics, Association for Computational Linguistics, pp. 2145–2158. Santa Fe, New Mexico, USA (2018). https://aclanthology.org/C18-1182

Yamada, I., Shindo, H., Takeda, H., Takefuji, Y.: Joint learning of the embedding of words and entities for named entity disambiguation. In: Proceedings of the 20th SIGNLL Conference on Computational Natural Language Learning, Association for Computational Linguistics, pp. 250–259. Berlin, Germany (2016). https://doi.org/10.18653/v1/K16-1025, https://aclanthology.org/K16-1025

Yang, Y., Chang, M.-W.: S-MART: novel tree-based structured learning algorithms applied to tweet entity linking. In: Proceedings of the 53rd Annual Meeting of the Association for Computational Linguistics and the 7th International Joint Conference on Natural Language Processing (Volume 1: Long Papers), Association for Computational Linguistics, pp. 504–513. Beijing, China (2015). https://doi.org/10.3115/v1/P15-1049, https://aclanthology.org/P15-1049

Yang, Y., Irsoy, O., Rahman, K.S.: Collective entity disambiguation with structured gradient tree boosting. In: Proceedings of the 2018 Conference of the North American Chapter of the Association for Computational Linguistics: Human Language Technologies, Volume 1 (Long Papers), Association for Computational Linguistics, pp. 777–786. New Orleans, Louisiana (2018). https://doi.org/10.18653/v1/N18-1071, https://aclanthology.org/N18-1071

Yarowsky, D., Ngai, G.: Inducing multilingual POS taggers and NP bracketers via robust projection across aligned corpora. In: Second Meeting of the North American Chapter of the Association for Computational Linguistics (2001). https://aclanthology.org/N01-1026

Yarowsky, D., Ngai, G., Wicentowski, R.: Inducing multilingual text analysis tools via robust projection across aligned corpora. In: Proceedings of the First International Conference on Human Language Technology Research (2001). https://aclanthology.org/H01-1035

Ye, Z., Ling, Z.-H.: Hybrid semi-Markov CRF for neural sequence labeling. In: Proceedings of the 56th Annual Meeting of the Association for Computational Linguistics (Volume 2: Short Papers), Association for Computational Linguistics, pp. 235–240. Melbourne, Australia (2018). https://doi.org/10.18653/v1/P18-2038, https://aclanthology.org/P18-2038

Yeniterzi, R.: Exploiting morphology in Turkish named entity recognition system. In: Proceedings of the ACL 2011 Student Session, Association for Computational Linguistics, pp. 105–110. Portland, OR, USA (2011). https://aclanthology.org/P11-3019

Yih, W.-T., Chang, M.-W., He, X., Gao, J.: Semantic parsing via staged query graph generation: Question answering with knowledge base. In: Proceedings of the 53rd Annual Meeting of the Association for Computational Linguistics and the 7th International Joint Conference on Natural Language Processing (Volume 1: Long Papers), Association for Computational Linguistics, pp. 1321–1331. Beijing, China (2015). https://doi.org/10.3115/v1/P15-1128, https://aclanthology.org/P15-1128

Zeman, D., Resnik, P.: Cross-language parser adaptation between related languages. In: Proceedings of the IJCNLP-08 Workshop on NLP for Less Privileged Languages (2008). https://aclanthology.org/I08-3008

Zhang, S., Qin, Y., Wen, J., Wang, X.: Word segmentation and named entity recognition for SIGHAN bakeoff3. In: Proceedings of the Fifth SIGHAN Workshop on Chinese Language Processing, Association for Computational Linguistics, pp. 158–161. Sydney, Australia (2006). https://aclanthology.org/W06-0126

Zhang, W., Su, J., Tan, C.L., Wang, W.T.: Entity linking leveraging automatically generated annotation. In: Proceedings of the 23rd International Conference on Computational Linguistics (Coling 2010), Coling 2010 Organizing Committee, pp. 1290–1298. Beijing, China (2010). https://aclanthology.org/C10-1145

Zhang, W., Sim, Y.-C., Su, J., Tan, C.-L.: Entity linking with effective acronym expansion, instance selection and topic modeling. In: Twenty-Second International Joint Conference on Artificial Intelligence (IJCAI) (2011)

Zhang, Y., Yang, J.: Chinese NER using lattice LSTM. In: Proceedings of the 56th Annual Meeting of the Association for Computational Linguistics (Volume 1: Long Papers), Association for Computational Linguistics, pp. 1554–1564. Melbourne, Australia (2018). https://doi.org/10.18653/v1/P18-1144, https://aclanthology.org/P18-1144

Zheng, Z., Li, F., Huang, M., Zhu, X.: Learning to link entities with knowledge base. In: Human Language Technologies: The 2010 Annual Conference of the North American Chapter of the Association for Computational Linguistics, pp. 483–491. Los Angeles, California (2010). https://aclanthology.org/N10-1072

Zhou, B., Khashabi, D., Tsai, C.-T., Roth, D.: Zero-shot open entity typing as type-compatible grounding. In: Proceedings of the 2018 Conference on Empirical Methods in Natural Language Processing, Association for Computational Linguistics, pp. 2065–2076. Brussels, Belgium (2018). https://doi.org/10.18653/v1/D18-1231, https://aclanthology.org/D18-1231

Zhou, S., Rijhwani, S., Wieting, J., Carbonell, J., Neubig, G.: Improving candidate generation for low-resource cross-lingual entity linking. Trans. Assoc. Comput. Linguistics **8**, 109–124 (2020). https://doi.org/10.1162/tacl_a_00303. https://aclanthology.org/2020.tacl-1.8

Zhu, Y., Wang, G.: CAN-NER: convolutional attention network for Chinese named entity recognition. In: Proceedings of the 2019 Conference of the North American Chapter of the Association for Computational Linguistics: Human Language Technologies, Volume 1 (Long and Short Papers), Association for Computational Linguistics, pp. 3384–3393. Minneapolis, Minnesota (2019). https://doi.org/10.18653/v1/N19-1342, https://aclanthology.org/N19-1342

Zhuo, J., Cao, Y., Zhu, J., Zhang, B., Nie, Z.: Segment-level sequence modeling using gated recursive semi-Markov conditional random fields. In: Proceedings of the 54th Annual Meeting of the Association for Computational Linguistics (Volume 1: Long Papers), Association for Computational Linguistics, pp. 1413–1423. Berlin, Germany (2016). https://doi.org/10.18653/v1/P16-1134, https://aclanthology.org/P16-1134

www.ingramcontent.com/pod-product-compliance
Lightning Source LLC
Chambersburg PA
CBHW050204310325
24335CB00006BA/258